GRAIL: Mapping the Moon's Interior

M.T. Zuber · C.T. Russell
Editors

GRAIL: Mapping the Moon's Interior

Previously published in *Space Science Reviews* Volume 178, Issue 1, 2013

 Springer

Editors

M.T. Zuber
Massachusetts Institute of Technology
Cambridge, MA, USA

C.T. Russell
University of California Los Angeles
Los Angeles, CA, USA

ISBN 978-1-4614-9583-3   ISBN 978-1-4614-9584-0 (eBook)
DOI 10.1007/978-1-4614-9584-0
Springer New York Heidelberg Dordrecht London

Library of Congress Control Number: 2013952929

Cover illustration: The figure shows the free-air gravity field of the Moon derived from data acquired by the twin GRAIL spacecraft that orbited the Moon between January 1 and December 17, 2012. The colors represent the magnitude of the gravity field where red is positive gravity and the blue is negative gravity with respect to the mean gravity field of the Moon. All craters greater than about 15 km in diameter are identifiable in the gravity field.

Printed on acid-free paper

Springer is part of Springer Science+Business Media (www.springer.com)

Contents

DOI 10.1007/978-1-4614-9584-0_1
Reprinted from *Space Science Reviews* Journal, DOI 10.1007/s11214-013-0010-x

Foreword

Maria T. Zuber · C.T. Russell

Received: 22 July 2013 / Accepted: 22 July 2013 / Published online: 1 August 2013
© Springer Science+Business Media Dordrecht 2013

In 1992 a small workshop in San Juan Capistrano marked the beginning of an innovation in planetary exploration, the Principal Investigator-led mission. NASA announced the establishment of a continuing "line item" in the budget for the development, launch and operation of missions led by a Principal Investigator from inside or outside NASA. These missions were to be less costly than flagship missions that addressed the major objectives of planetary exploration. They would be more focused, developed more quickly for flight, with a limited number of instruments and a limited number of investigators. They would ensure that the smaller but important objectives of the planetary program would be addressed. The first two missions were selected in a mode similar to the earlier selection process to get the program off to a quick start but soon a new process was established. The best mission or pair of missions was to be selected from a group of about thirty proposals. From this process arose missions approved to go to the Moon, bring back solar wind and comet samples, to excavate a crater on a comet, to orbit Mercury, to orbit main belt asteroids, and to identify Earth-like exoplanets.

There is a rule of thumb in planetary exploration that every time you increase your resolution an order of magnitude, you make major discoveries. The increase in resolution is one of the reasons that missions to orbit planetary bodies are so important. You cannot obtain the needed resolution with telescopes at 1AU and a planetary flyby gives high resolution data only briefly and often on only one side of the body visited. One might not have thought that the Moon was ripe for such a revolutionary advance having been orbited closely in preparation for and during the Apollo program and visited multiple times in recent years but in one area it certainly was.

The GRACE mission in Earth orbit had shown the power of differential gravity measurements using a pair of spacecraft whose relative positions were accurately known. The

M.T. Zuber
Massachusetts Institute of Technology, Cambridge, MA 02139-4307, USA

C.T. Russell (✉)
University of California Los Angeles, Los Angeles, CA 90095, USA
e-mail: ctrussel@igpp.ucla.edu

team that successfully proposed the GRAIL mission knew that such measurements were also achievable on the Moon and could advance our understanding immeasurably at that body. This relatively brief two spacecraft mission has now been completed and the concept was a complete success. The Moon's gravitational field is now measured far more accurately and at higher resolution than ever before, and in fact better than the Earth. Although the data analysis is still in early stages the geophysics of the Moon is now far better understood and many paradigms have been turned on end.

Like the mission this special issue is short and to the point. It contains three papers from the mission team. The first paper by Maria Zuber and colleagues, provides an overview of the GRAIL mission; the science objectives, measurements, spacecraft, instruments and mission development, design, data flow and products. The second paper by S. Asmar and colleagues describes the scientific measurement system of the mission including the early modeling and simulation efforts. These enabled the scientific requirements to be converted to engineering specifications that became the primary drivers for development and testing. The third paper describes the implementation, testing and performance of the instrument complement flown on the two spacecraft.

The successful implementation of a mission as sophisticated as the Gravity Recovery and Interior Laboratory requires the hard work and assistance of many talented and dedicated individuals. In this instance these are not just the scientists involved, many of whom are the authors and coauthors of these articles, but also the engineers at the Jet Propulsion Laboratory and Lockheed-Martin Space Systems Company who designed, implemented, tested, and integrated the two spacecraft and their payloads, as well as the management at NASA and all who contributed to this mission. We are grateful to them all. The success of this volume is also due to many people. First of all the editors wish to thank the authors who distilled the voluminous material mission development produces, into highly readable articles. The editors also benefited from an excellent set of referees who acted as a test audience and helped refine the manuscript provided by the authors. These referees included Glenn Cunningham, Cheryl Gramling, Walter S. Kiefer, Ryan S. Park, Byron D. Tapley, Slava Turyshev.

Equally important has been the strong support this project received from Harry Blom, Jennifer Satten, Esther Rentmeester and Lalitha Jaganathan at Springer. At UCLA we were skillfully assisted by Marjorie Sowmendran who acted as the interface between the authors, referees and the publisher.

DOI 10.1007/978-1-4614-9584-0_2
Reprinted from *Space Science Reviews* Journal, DOI 10.1007/s11214-012-9952-7

Gravity Recovery and Interior Laboratory (GRAIL): Mapping the Lunar Interior from Crust to Core

Maria T. Zuber · David E. Smith · David H. Lehman ·
Tom L. Hoffman · Sami W. Asmar · Michael M. Watkins

Received: 1 September 2012 / Accepted: 26 November 2012 / Published online: 4 January 2013
© Springer Science+Business Media Dordrecht 2012

Abstract The Gravity Recovery and Interior Laboratory (GRAIL) is a spacecraft-to-spacecraft tracking mission that was developed to map the structure of the lunar interior by producing a detailed map of the gravity field. The resulting model of the interior will be used to address outstanding questions regarding the Moon's thermal evolution, and will be applicable more generally to the evolution of all terrestrial planets. Each GRAIL orbiter contains a Lunar Gravity Ranging System instrument that conducts dual-one-way ranging measurements to measure precisely the relative motion between them, which in turn are used to develop the lunar gravity field map. Each orbiter also carries an Education/Public Outreach payload, Moon Knowledge Acquired by Middle-School Students (MoonKAM), in which middle school students target images of the Moon for subsequent classroom analysis. Subsequent to a successful launch on September 10, 2011, the twin GRAIL orbiters embarked on independent trajectories on a 3.5-month-long cruise to the Moon via the EL-1 Lagrange point. The spacecraft were inserted into polar orbits on December 31, 2011 and January 1, 2012. After a succession of 19 maneuvers the two orbiters settled into precision formation to begin science operations in March 1, 2012 with an average altitude of 55 km. The Primary Mission, which consisted of three 27.3-day mapping cycles, was successfully completed in June 2012. The extended mission will permit a second three-month mapping phase at an average altitude of 23 km. This paper provides an overview of the mission: science objectives and measurements, spacecraft and instruments, mission development and design, and data flow and data products.

Keywords Gravity · Moon · Lunar · Remote sensing · Spacecraft

M.T. Zuber (✉) · D.E. Smith
Department of Earth, Atmospheric and Planetary Sciences, Massachusetts Institute of Technology,
Cambridge, MA 02139-4307, USA
e-mail: zuber@mit.edu

D.H. Lehman · T.L. Hoffman · S.W. Asmar · M.M. Watkins
Jet Propulsion Laboratory, California Institute of Technology, Pasadena, CA 91109, USA

1 Introduction

In December 2007, NASA competitively selected the Gravity Recovery and Interior Laboratory (GRAIL) mission under the Solar System Exploration Division Discovery Program. GRAIL was developed to map the structure of the lunar interior from crust to core. This objective will be accomplished by producing detailed maps of the lunar gravity field at unprecedented resolution. These gravity maps will be interpreted in the context of other observations of the Moon's interior and surface obtained by orbital remote sensing and surface samples, as well as experimental measurements of planetary materials. The resulting improved knowledge of the interior will be used to understand the Moon's thermal evolution, and by comparative planetological analysis, the evolution of other terrestrial planets. GRAIL is unique in that it provides a focused measurement to address broad scientific objectives.

The GRAIL mission is led by the Massachusetts Institute of Technology. The project is managed by the Jet Propulsion Laboratory (JPL), with Lockheed-Martin Space Systems Corporation (LMSSC) contracted to provide the spacecraft. GRAIL's science instrument was developed by JPL. Education and Outreach is implemented by Sally Ride Science. The Science Team contains representation from 15 academic institutions and NASA Centers.

After a successful launch on September 10, 2011 and a 3.5-month-long trans-lunar cruise, the twin GRAIL orbiters, named Ebb and Flow, were placed into a polar orbit on December 31, 2011 and January 1, 2012. After a succession of 19 maneuvers the two orbiters settled into a precision formation to begin science operations a week earlier than planned, on March 1, 2012, at an average altitude of 55 km. The Primary Mission (PM) was completed on May 29, 2012. On the basis of a competitive proposal evaluation, NASA decided to extend the GRAIL mission until December 2012.

Each GRAIL orbiter contains a Lunar Gravity Ranging System (LGRS) (Klipstein et al. 2012) instrument that conducts dual-one-way ranging to precisely measure the relative motion between the two spacecraft. These distance changes are used to develop the lunar gravity field map (Thomas 1999). The LGRS is a modified version of an instrument used on the Gravity Recovery and Climate Experiment (GRACE) (Tapley et al. 2004) mission which is currently gravity mapping the Earth. GRAIL's twin spacecraft have heritage derived from an experimental U.S. Air Force satellite (XSS-11) and the Mars Reconnaissance Orbiter (MRO) mission (Johnson et al. 2005), both developed by LMSSC. Each orbiter carries an Education/Public Outreach (E/PO) payload called GRAIL MoonKAM (or Moon Knowledge Acquired by Middle-School Students) in which middle school students target images of the Moon.

In this paper, Sect. 2 motivates study of the Moon's interior and describes previous attempts to measure the gravity field; Sect. 3 summarizes the GRAIL science objectives in the context of outstanding questions in lunar science; Sect. 4 summarizes the spacecraft and instruments; Sect. 5 covers the Mission Development and Design; Sect. 6 describes GRAIL's extended mission; Sect. 7 describes data flow; and Sect. 8 describes GRAIL's data products. All acronyms are defined in the Appendix.

2 The Lunar Interior and the Measurement of Planetary Gravity

The Moon is the most accessible and best studied of the rocky (a.k.a. "terrestrial") planetary bodies beyond Earth. Unlike Earth, the Moon's surface geology preserves the record of nearly the entirety of 4.5 billion years of solar system history. Orbital observations combined with samples of surface rocks returned to Earth from known locations make the Moon unique in providing a detailed, global record of the geological history of a terrestrial planetary body, particularly the early history subsequent to accretion.

The structure and composition of the lunar interior (and by inference the nature and timing of compositional differentiation and of internal dynamics) hold the key to reconstructing this history. For example, longstanding questions such as the origin of the maria, the reason for the nearside-farside asymmetry in crustal thickness, the role of mantle dynamics in lunar thermal evolution, and the explanation for the puzzling magnetization of crustal rocks, all require a greatly improved understanding of the Moon's interior.

Moreover, deciphering the structure of the interior will bring understanding of the evolution of not only the Moon itself, but also of the other terrestrial planets (Paulikas et al. 2007). For example, while the Moon was once thought to be unique in developing a "magma ocean" shortly after accretion (Wood et al. 1970), such a phenomenon has now been credibly proposed for Mars as well (Elkins-Tanton 2008). Insight into fundamental processes such as the role of impacts in perturbing internal thermal state and in the re-distribution of crust are relevant to all solid planetary bodies.

Gravity is the primary means of mapping the mass distribution of the interior, but the Moon presents a special challenge in sampling the global field. A spacecraft in orbit is perturbed by the distribution of mass at the surface and within a planetary body, particularly that beneath the spacecraft as it orbits overhead. The measurement of planetary gravity has most commonly been achieved by monitoring the frequency shift of a spacecraft's radio signal measured in the line of sight between the spacecraft, while in orbit about a planetary body, and a tracking station on Earth (Phillips et al. 1978). The Doppler shift of the radio frequency provides a measure of spacecraft velocity, which when differenced provides accelerations. Correcting for accelerations due to spacecraft thrusting and maneuvering as well as other non-gravitational forces (Asmar et al. 2012) yields the gravitational field of the planet. For this approach to work, all parts of the planetary surface must be visible in the line-of-sight of the ground station as the planet rotates beneath the spacecraft. However, because the Moon is in synchronous rotation about Earth the farside is never directly visible; thus gravity on the nearside is sensed much more accurately than on the farside.

The Moon was the first planetary body beyond Earth for which gravity field information was obtained with a spacecraft, beginning with the Russian Luna 10 (Akim 1966). Subsequent U.S. efforts included Lunar Orbiters 1–5, the Deep Space Program Science Experiment (DSPSE; Clementine) (Zuber et al. 1994; Lemoine et al. 1997) and Lunar Prospector (Konopliv et al. 1998, 2001).

The geometrical shortcoming associated with lack of visibility from Earth of a spacecraft over the farside motivated the use of sub-satellites in the recent Kaguya mission (Namiki et al. 2009). A sub-satellite can be tracked by the orbiter on the farside to measure gravitational perturbations when not in the line of sight from Earth.

The line-of-sight method produced reasonable measurement of gravity of the Moon's nearside that most famously led to the early identification of "mascons" (Muller and Sjogren 1968; Phillips et al. 1972), lunar mass concentrations spatially associated with the Moon's mare basins. Other analyses developed local gravity representations using surface mass models (Wong et al. 1971; Ananda 1977). The most natural representation of the gravity field is a spherical harmonic expansion, since spherical harmonics are the solution to Laplace's equation, $\nabla^2 U = 0$, which describes the gravitational potential U, on a sphere. The spherical harmonic solution for the gravitational potential with normalized coefficients $(\bar{C}_{nm}, \bar{S}_{nm})$ can be expressed (Kaula 1966; Heiskanen and Moritz 1967)

$$U = \frac{GM}{r} + \frac{GM}{r} \sum_{n=2}^{\infty} \sum_{m=0}^{n} \left(\frac{R_e}{r}\right)^n \bar{P}_{nm}(\sin\phi_{lat})\left[\bar{C}_{nm}\cos(m\lambda) + \bar{S}_{nm}\sin(m\lambda)\right], \quad (1)$$

Table 1 Summary of recent lunar gravity models

Reference	Field	Data used	Spherical harmonic degree and order
Lemoine et al. (1997)	GLGM-1	Lunar Orbiter 1–5, Apollo subsatellites, Clementine	70 × 70 (78-km blocksize)
Konopliv et al. (2001)	LP100, LP150	Lunar Orbiter 1–5, Apollo sub-satellites, Clementine, Lunar Prospector	100 × 100 (54-km blocksize) later updated to 150 × 150 (36-km blocksize); Useful for geophysical modeling to 70 × 70 (78-km blocksize)
Matsumoto et al. (2010)	SGM100h	Kaguya S-band and X-band; Orbiter, Relay subsatellite, VLBI subsatellite	100 × 100 (54-km blocksize); Useful for global geophysics to 70 × 70 (78-km blocksize)
Mazarico et al. (2010)	GLGM-3	Lunar Orbiter 1–5, Apollo sub-satellites, Clementine, Lunar Prospector	150 × 150 (36-km blocksize)
GRAIL		Satellite-to-satellite tracking (Ka-band); X-band link to Earth	At least 180 × 180 (30-km blocksize) expected

where GM is the gravitational constant times the mass of the Moon, n is the degree, m is the order, $\bar{P}_{nm}$ are the fully normalized associated Legendre polynomials, R_e is the reference radius of the Moon, ϕ_{lat} is the latitude, and λ is the longitude (east positive). The gravity coefficients are normalized and are related to the unnormalized coefficients according to (Kaula 1966)

$$\begin{pmatrix} C_{nm} \\ S_{nm} \end{pmatrix} = \left[\frac{(n-m)!(2n+1)(2-\delta_{0m})}{(n+m)!} \right]^{1/2} \begin{pmatrix} \bar{C}_{nm} \\ \bar{S}_{nm} \end{pmatrix} = f_{nm} \begin{pmatrix} \bar{C}_{nm} \\ \bar{S}_{nm} \end{pmatrix}. \tag{2}$$

The coefficients contain the information about the variation of gravity, and n and m describe the resolution of the field, which in practice is dictated by coverage and spacecraft altitude. A comparison of recent spherical harmonic solutions for the lunar gravitational field to that expected from GRAIL is given in Table 1. A companion paper (Asmar et al. 2012) discusses extensive simulations that assessed the expected quality of the GRAIL field on the basis of

Table 2 Primary Mission science investigations

Science objective	Science investigation	Area (10^6 km^2)	Resolution (km)	Requirements (30-km block)
Determine the structure of the lunar interior	1. Crust & Lithosphere	~10	30	±10 mGal
	2. Thermal Evolution	~4	30	±2 mGal
	3. Impact Basins	~1	30	±0.5 mGal
	4. Magmatism	~0.1	30	±0.1 mGal
Advance understanding of the thermal evolution of the Moon	5. Deep Interior	N/A	N/A	$k_2 \pm 6 \times 10^{-4}$ (3 %)
	6. Inner Core Detection	N/A	N/A	$k_2 \pm 2.2 \times 10^{-4}$ (1 %) $C_{2,1} \pm 1 \times 10^{-10}$

quantitative assessment of various deterministic and stochastic errors on the measurements and on the recovery of the gravitational field.

3 Primary Mission Science Objectives

The necessity to understand the Moon's internal structure in order to reconstruct planetary evolution motivates the GRAIL primary science objectives, which are to:

- Determine the structure of the lunar interior from crust to core, and
- Advance understanding of the thermal evolution of the Moon.

In addition GRAIL has one secondary objective:

- Extend knowledge gained on the internal structure and thermal evolution of the Moon to other terrestrial planets.

The primary objectives are closely related; interior structure along with surface geology and chemistry are required to reconstruct thermal evolution. Of these, it is knowledge of the internal structure that is currently most lacking (Hood and Zuber 2000). The secondary objective adds a comparative planetological focus to the mission and affords the opportunity to engage a broader cross section of the scientific community with expertise in terrestrial planet evolution.

GRAIL's Primary Mission includes six lunar science investigations, to:

1. Map the structure of the crust and lithosphere.
2. Understand the Moon's asymmetric thermal evolution.
3. Determine the subsurface structure of impact basins and the origin of mascons.
4. Ascertain the temporal evolution of crustal brecciation and magmatism.
5. Constrain the deep interior structure from tides.
6. Place limits on the size of the possible inner core.

Measurement requirements for GRAIL's science investigations are given in Table 2 and Fig. 1. The GRAIL Science Team, listed in Table 3, carries out the science investigations.

The mission accomplishes its broad lunar science objectives via a focused, extremely precise measurement: the distance change between two spacecraft. Specifically, GRAIL obtains global, regional and local high-resolution (30 × 30-km), high-accuracy (<10-mGal) gravity field measurements with twin, low-altitude (55 km) polar-orbiting spacecraft.

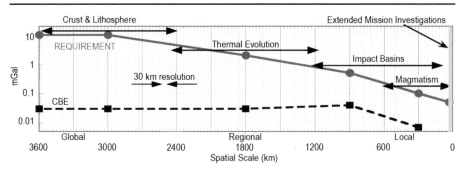

Fig. 1 Primary Mission science measurement performance and requirements. CBE refers to the "current best estimate" of GRAIL Primary Mission performance prior to launch

4 Spacecraft and Instruments

The GRAIL orbiters (Hoffman 2009) are nearly identical spacecraft with heritage to past spacecraft and spacecraft subsystems built at LMSSC. The main structure and propulsion system are based on the design used on the U.S. Air Force Experimental Satellite System 11 (XSS-11), built under contract by LMSSC, that was intended to demonstrate autonomous rendezvous and proximity maneuvers. Some components of the GRAIL spacecraft, most notably the flight computer, traced heritage to the Mars Reconnaissance Orbiter (MRO), built by LMSSC under contract to the Jet Propulsion Laboratory. While MRO was a larger spacecraft with numerous instruments and complex targeting requirements, the basic processing functions from MRO were transferable to GRAIL. Functionally, the GRAIL flight computer is a simplification of that on MRO, inspiring the moniker "MRO-Lite". Both XSS-11 and MRO were successful projects and collectively formed a sound basis for the design heritage of the GRAIL mission. The two orbiters were designed to be as identical as possible to reduce cost, eliminate configuration complexities and streamline integration and test flows. Small differences in design were necessitated by geometrical constraints associated with satellite-to-satellite ranging. Two views of the spacecraft are shown in Figs. 2a and 2b.

A key attribute of the orbiters is that they are single string for almost every component. A single-string mission allows for a much more simplified system in terms of design, and integration, test and operations, but it entails increased risk. The approach was made to fit with the GRAIL mission's guiding principle of "low risk implementation" due to the relatively short primary mission life (9 months) coupled with the adoption of a robust reliability program (Taylor et al. 2012). The single string approach also reduced the overall development cost of the mission, and in addition minimized mass, allowing both spacecraft to be launched on a Delta-II Heavy with considerable mass margin. The only exceptions to the single-string philosophy were areas where redundancy either came at very low cost or where the absence of redundancy would pose a risk to the project (Hoffman 2009).

There are two payloads on each GRAIL orbiter. The science instrument is the Lunar Gravity Ranging System (LGRS) (Klipstein et al. 2012) and the Education/Public Outreach (E/PO) payload is Moon Knowledge Acquired by Middle-School Students (MoonKAM). GRAIL is unique in that the science for the mission is achieved with a single instrument. Each of the payloads is briefly described herein, and detailed discussion of the LGRS is given by Klipstein et al. (2012).

The LGRS instrument utilizes a dual-one-way ranging (DOWR) measurement to precisely measure the relative motion between the two orbiters. The fundamental method used

Table 3 GRAIL science team

Team member	Role	Institution
Maria T. Zuber	Principal Investigator	Massachusetts Institute of Technology
David E. Smith	Deputy Principal Investigator	Massachusetts Institute of Technology
Michael M. Watkins	Co-Investigator/Project Scientist	Jet Propulsion Laboratory
Sami W. Asmar	Co-Investigator/Project Scientist	Jet Propulsion Laboratory
Alexander S. Konopliv	Co-Investigator	Jet Propulsion Laboratory
Frank G. Lemoine	Co-Investigator	NASA/Goddard Space Flight Center
H. Jay Melosh	Co-Investigator	Purdue University
Gregory A. Neumann	Co-Investigator	NASA/Goddard Space Flight Center
Roger J. Phillips	Co-Investigator	Southwest Research Institute
Sean C. Solomon	Co-Investigator	Lamont-Doherty Earth Observatory of Columbia University
Mark A. Wieczorek	Co-Investigator	Institute de Physique du Globe de Paris
James G. Williams	Co-Investigator	Jet Propulsion Laboratory
Jeffrey Andrews-Hanna	Guest Scientist	Colorado School of Mines
James Head	Guest Scientist	Brown University
Walter Kiefer	Guest Scientist	Lunar and Planetary Institute
Isamu Matsuyama	Guest Scientist	University of Arizona
Patrick McGovern	Guest Scientist	Lunar and Planetary Institute
Francis Nimmo	Guest Scientist	University of California, Santa Cruz
Christopher Stubbs	Guest Scientist	Harvard University
G. Jeffrey Taylor	Guest Scientist	University of Hawaii, Honolulu
Renee Weber	Guest Scientist	NASA/Marshall Space Flight Center

for the ranging measurement has a long history of use over dozens of missions as a primary navigation tool. This method was extended on the GRACE mission (Dunn et al. 2002), which has been successfully mapping the Earth's gravity field since launch in March, 2002 (Tapley et al. 2004).

As shown in the block diagram in Fig. 3, the instrument consists of a Ka-band antenna for transmitting and receiving inter-satellite signals; a microwave assembly (MWA) for generating Ka-band signals for transmission and mixing down the inter-satellite signals; a Gravity Recovery Processor Assembly (GPA) for processing both the Ka-band signals and those from the S-band Time Transfer System (TTS), the latter of which is used to correlate inter-

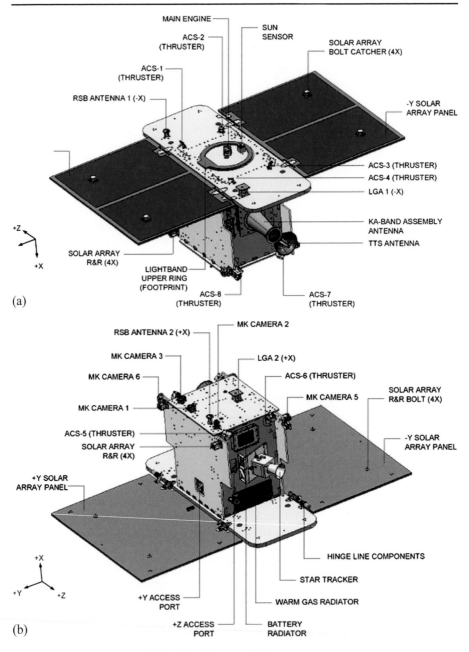

Fig. 2 GRAIL (**a**) −X spacecraft view; (**b**) +X spacecraft view

satellite ranges; an Ultra-Stable Oscillator (USO) that provides timing for both the Ka-band and S-band systems; and a Radio Science Beacon (RSB) that provides an X-band Doppler carrier to support daily calibration of the USO frequency by the DSN. The elements of the instrument work together to achieve micron-level precision relative range differences.

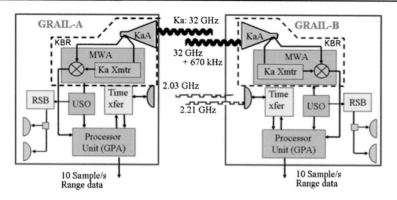

Fig. 3 Lunar Gravity Ranging System block diagram

Overall science implementation is achieved through the LGRS instrument measuring the change in range between the two orbiters. The gravity field of the Moon influences the motion of the center-of-mass (CM) of each spacecraft, which essentially acts like a proof mass in orbit about the Moon. Surface features such as craters, mass concentrations (mascons), and deep interior structure and dynamics perturb the spacecraft orbits and introduce variations in the relative motion between the spacecraft (Thomas 1999).

The fundamental measurement is the Line of Sight (LOS) range rate, which was designed to achieve an accuracy of 4.5 $\mu m\,s^{-1}$ over a 5-second sample interval. These data are collected along with DSN tracking data over a period of 27.3 days, providing global coverage of the Moon, six times over. The entire set of mapping cycle data is then processed to recover the global gravity map.

The MoonKAM payload is a set of cameras, designed and built by Ecliptic Enterprises Corporation, that image the lunar surface. The MoonKAM investigation was led by America's first woman in space, Dr. Sally K. Ride, and since her untimely passing in July 2012 continues to be implemented by Sally Ride Science (SRS). MoonKAM is the first planetary imaging experiment dedicated entirely to education and outreach. GRAIL's MoonKAMs consist of electronics and four camera heads per spacecraft to allow imaging at a variety of directions and resolutions.

The MoonKAM investigation is targeted at the middle school level but accepts participation from all supervised student groups and clubs. At the end of the PM the program had enlisted over 2800 participating classrooms and/or student organizations and over 100,000 individual participants. Students use trajectory software based on JPL's "Eyes on the Solar System" (http://solarsystem.nasa.gov/eyes/) to target images of the lunar surface. The cameras are operated by undergraduates at the University of California, San Diego, who are supervised by personnel at SRS. All acquired images are posted to a public website and classroom activities developed by staff at Sally Ride Science are available for subsequent scientific analysis of the lunar surface. The MoonKAM investigation is intended to motivate interest in science, technology and mathematics by providing meaningful, early exposure to the challenges and processes used in spacecraft operations, and genuine participation in exploration and scientific analysis.

Table 4 Summary of key GRAIL events

Event	Date
GRAIL Selection as a Discovery Mission	Dec. 2007
Preliminary Design Review	Nov. 2008
Confirmation Review	Jan. 2009
Critical Design Review	Nov. 2009
Systems Integration Review	June 2010
Pre-Ship Review	May 2011
Orbiters delivered to launch site	May 2011
Launch	Sept. 10, 2011
Lunar Orbit Insertion	Dec. 31, 2011 (GR-A)/Jan. 1, 2012 (GR-B)
Primary Science Phase Begins	March 2012
Primary Science Phase Ends	June 2012
Extended Science Phase Ends	December 2012
Archive of Prime Mission Levels 0 and 1 data	December 2012
Archive of Prime Mission Levels 2 and 3 data	September 2013
Archive of Extended Mission Levels 0 and 1 data	June 2013
Archive of Extended Mission Levels 2 and 3 data	June 2014

5 Mission Development and Design

A summary of key dates in the development of the GRAIL mission is given in Table 4. GRAIL's development initiated subsequent to competitive selection as NASA's 11th Discovery mission in December 2007. During its development phase, GRAIL met all milestones on time, including reviews, and the delivery of gate products and the delivery of NASA life-cycle prescribed documentation and development products, and hardware. The LGRS instrument was delivered for integration to LMSSC 2 weeks early. In addition, the Project team compressed the Assembly and Test schedule to deliver both spacecraft to the launch facility a week early (Taylor et al. 2012). This early delivery was in support of a risk reduction request from the launch vehicle team at NASA Kennedy Space Center to allow additional processing time due GRAIL's status as the last east coast Delta launch. GRAIL's development concluded with a successful launch, on schedule and under budget (GAO 2012), of the twin spacecraft on a single Delta-II 7920H rocket from the Cape Canaveral Air Force Station (CCAFS) in Florida on September 10, 2011.

The GRAIL orbiters must fly in precise formation to map the Moon while at the same time pointing their body-fixed solar panels toward the Sun. Because of the importance of the sun's direction, a parameter of particular relevance in the GRAIL mission design is the solar beta angle (β), defined as the angle between the orbital plane and a line drawn from the Sun to the Moon. Over the course of a year, the position of the sun with respect to the orbit plane dictates the times and duration of periods when the solar panels receive enough sunlight to power the spacecraft and perform science operations. The GRAIL orbiters, which carried small onboard batteries (Hoffman 2009), were designed to map at solar beta angles $\beta > 49°$ but in practice nominal operations were possible for $\beta > 40°$, which enabled science mapping to initiate a week early in the PM. It is convenient and informative to graphically depict GRAIL's mission phases from a heliocentric perspective, presented in Fig. 4 (Roncoli and Fujii 2010).

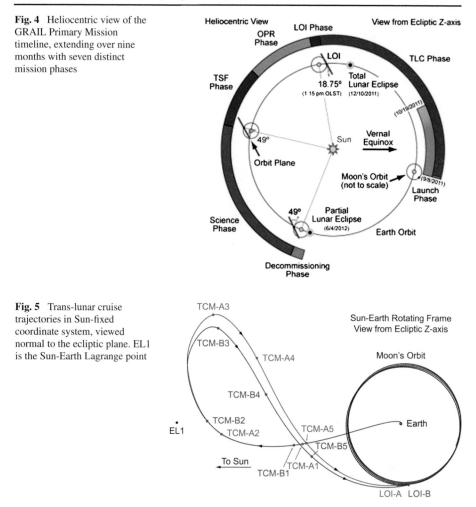

Fig. 4 Heliocentric view of the GRAIL Primary Mission timeline, extending over nine months with seven distinct mission phases

Fig. 5 Trans-lunar cruise trajectories in Sun-fixed coordinate system, viewed normal to the ecliptic plane. EL1 is the Sun-Earth Lagrange point

Following launch, the Delta upper stage that contained both spacecraft entered a parking orbit and then injected into the trans-lunar trajectory, initiating the Trans-Lunar Cruise (TLC) phase of the mission (Chung et al. 2010). Subsequent to injection toward the Moon, the two spacecraft were deployed from the launch vehicle and traveled to the Moon upon similar but separate trajectories. As shown schematically in Fig. 5, the TLC phase (Chung et al. 2010) utilized a 3.5-month, low-energy trajectory via the Sun-Earth Lagrange point (EL-1) to transit to the Moon. This unique mission design (Roncoli and Fujii 2010) provides several key features important to the GRAIL mission. First, the low energy trajectory allowed for an extended launch window, providing a 42-day window versus a 3–6-day window for a direct trajectory. Second, the low-energy trajectory allowed for a smaller required delta-V for lunar orbit insertion that in turn allowed for smaller propulsion system on the spacecraft; this prevented a large-scale redesign of the heritage spacecraft. Third, this particular mission design allowed for a fixed Lunar Orbit Insertion phase for any date within the 42-day launch window, which allowed planning for cruise operations to be decoupled from orbital operations; an additional benefit is that the orbit insertions were able to be separated by 25 hours to avoid the execution of two critical events in a single day. Finally, the

Fig. 6 Schematic of Orbit Period Reduction phase as viewed from the Moon's north pole (*top*) and from Earth (*bottom*). Also shown at bottom in gold is the LOI burn arc for GRAIL-A (Ebb)

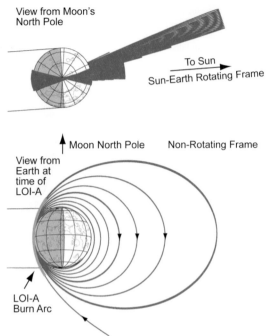

TLC period allowed time to perform spacecraft and payload checkout, allowed time for the Ultra-Stable Oscillator (USO) to stabilize, and allowed the spacecraft to outgas. Outgassing is a non-conservative force that could influence gravity measurements if not done prior to gravity mapping.

The first Lunar Orbit Insertion (LOI) maneuver (Hatch et al. 2010) occurred on December 31, 2011. And the second occurred on January 1, 2012. These maneuvers involved 39-minute-long continuous main engine propulsive burns of ~190 m/s to slow the spacecraft sufficiently to enter lunar capture orbits. The LOI burns were conducted so as to allow for continuous command and telemetry coverage from the NASA Deep Space Network (DSN).

After LOI, the two spacecraft underwent orbit circularization and were positioned into formation to prepare for science operation. Orbit circularization took approximately one month during the mission's Orbit Period Reduction (OPR) phase, which is shown schematically in Fig. 6. The main activity during this phase was to perform Period Reduction Maneuvers (PRMs) (Hatch et al. 2010), that were designed to place each spacecraft into a 55-km altitude circular orbit with the approximate desired separation and formation required for science. After OPR, the dual spacecraft went through a month-long Transition to Science Formation (TSF) phase during which a series of maneuvers established the proper formation and separation between the two spacecraft prior to the start of science collection. A total of 19 maneuvers following LOI were required in a two-month period prior to the start of science collection. However, during PM science mapping only one burn was executed, to adjust the drift rate between the two spacecraft. Figure 7 shows the actual variation of periapsis and apoapsis during the PM and Fig. 8 shows the corresponding evolution of distance and drift rate between the spacecraft.

During the 89-day Science phase, the GRAIL spacecraft completed over three 27.3-day mapping cycles (lunar sidereal periods) of the Moon and returned 637 Mbytes of science volume or >99.99 % of possible data. In performing its science mission, GRAIL achieved the

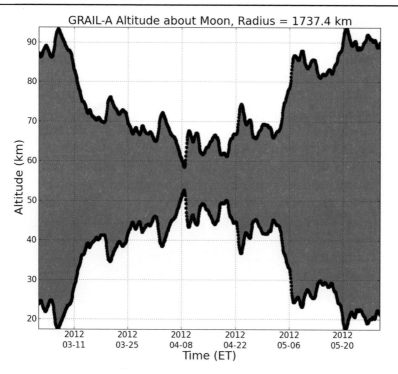

Fig. 7 Periapsis and apoapsis altitudes in the Primary Mission

first robotic demonstration of precision formation flying around another planetary body besides Earth (Roncoli and Fujii 2010). The PM ended with maneuvers to raise the spacecraft orbits on May 29, 2012. From launch through the PM, a total of 28 spacecraft maneuvers were performed, all flawlessly, by the GRAIL operations team.

6 Extended Mission

On the basis of competitive review, NASA has approved an Extended Mission (XM) for GRAIL, through December 2012, or approximately three months of data acquisition, that will enable collection of higher-resolution gravity data by flying the dual spacecraft in formation at an even lower altitude. The XM dramatically expands the scope of GRAIL's gravity science investigation beyond what was possible during the PM. By operating the dual spacecraft in the lowest orbit the flight team can safely support, it increases the resolution of the gravity field measurement by over a factor of two, sufficient to distinguish gravitational features down to a fraction of the crustal thickness. We defined "safely support" to mean that a missed or off-nominal maneuver could be recovered from in less than a week. Thus, GRAIL's XM enters a new realm of lunar science: crustal geophysics at the spatial scale of regional geology. It provides a singular opportunity to globally map the detailed structure of a planetary crust.

A heliocentric view of the GRAIL XM is shown in Fig. 9 (Sweetser et al. 2012). In late May 2012, when the PM was completed, periapsis raise maneuvers circularized the spacecraft orbits at an altitude of ~84 km for the low-activity Low Beta Angle phase. For

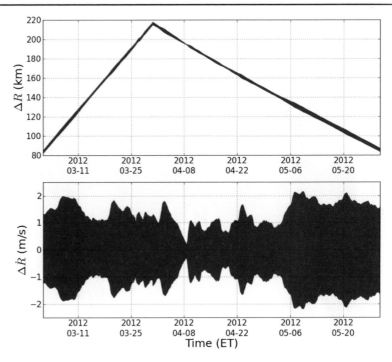

Fig. 8 Spacecraft (*top*) separation distance and (*bottom*) drift rate during the Primary Mission

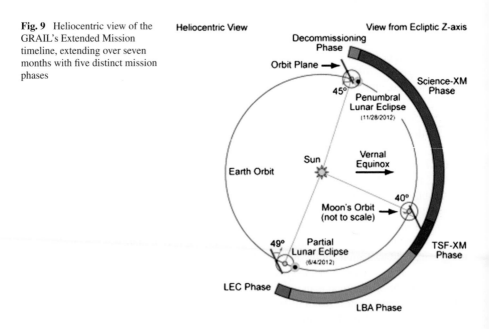

Fig. 9 Heliocentric view of the GRAIL's Extended Mission timeline, extending over seven months with five distinct mission phases

ten weeks subsequent to the lunar eclipse passage on June 4, the orientation of the orbit plane relative to the Sun did not allow for operation of the LGRS payloads while in orbiter-point configuration due to the Sun-Moon-Earth geometry. A second three-month science phase

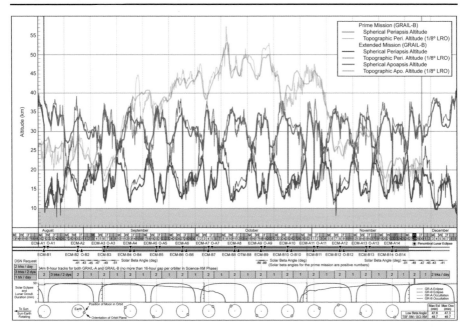

Fig. 10 Maximum (*red*) and minimum (*blue*) altitudes of GRAIL in the Extended Mission. *Thick lines* are altitudes with respect to a sphere of radius 1737.4 km and *thin lines* are with respect to LRO/LOLA topography. For comparison, *light green lines* show the periapsis altitudes in the Primary Mission

initiated successfully on August 30 when the solar beta angle reached 40°, at which time the solar panels were oriented in a manner that allows them to adequately charge the spacecraft while in ranging configuration.

GRAIL's XM average altitude is 23-km, less than half the average altitude of the PM. Because of the low orbital altitude, operations in the Extended Mission are far more complex than in the PM (Wallace et al. 2012). Unlike the PM, which featured only one thrust maneuver to change the drift rate of the spacecraft over three months of mapping, the XM requires three maneuvers a week to maintain the mapping altitude (Sweetser et al. 2012). During extended science mapping, weekly eccentricity correction maneuvers (ECMs) on both orbiters maintain the orbits. There will also be a weekly Orbit Trim Maneuver (OTM) on one orbiter, a day after the ECMs, to control the orbiter separation distance. The XM contains 46 baseline maneuvers. The altitude variation about the mean will be constrained to ±12 km. Figure 10 shows the variation in altitudes of the spacecraft during the XM. The XM mapping orbit is deemed by analysis (Wallace et al. 2012) to be the lowest orbit the flight team can safely support; the resulting resolution of the gravity field measurements will correspond to spatial blocksizes from ~30 km to <10 km. In addition, the accuracy will be improved by over an order of magnitude at 30-km block size resolution, permitting GRAIL to map structure globally within the upper crust.

The orbiters will experience a penumbral lunar eclipse on November 28. Subsequent to this event the altitude of the orbiters will again be decreased, to an average altitude of 11 km, for additional mapping in the mission's "limbo phase". High-resolution of the Orientale Basin, the youngest large impact basin on the Moon, is planned for this period. Science mapping will end on December 14, after which a series of engineering experiments is planned

Table 5 Extended Mission science investigations

Investigation	Spatial scale & accuracy reqt
Structure of impact craters	12 km, 0.02 mGal
Near-surface magmatism	30 km, 0.01 mGal
Mechanisms and timing of deformation	12 km, 0.005 mGal
Cause(s) of crustal magnetization	12 km, 0.002 mGal
Estimation of upper crustal density	12 km, 0.005 mGal
Mass bounds on polar volatiles*	30 km, 0.002 mGal

* Assumes a 10-m-thick layer composed of 5 % H_2O ice, 95 % regolith

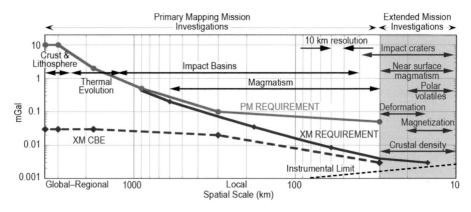

Fig. 11 Extended Mission science measurement performance and requirements. The current best estimate (CBE) has significant margin over the requirements. The demonstrated instrument performance during test in cruise was 0.001 mGal at 1 Hz

prior to deorbit on December 17, 2012. The mission end game includes a burn to depletion maneuver for each spacecraft.

The GRAIL XM has one overarching science objective:

- Determine the structure of lunar highland crust and maria, addressing impact, magmatic, tectonic and volatile processes that have shaped the near surface.

To address this objective the GRAIL extended mission undertakes six investigations:

1. Structure of impact craters.
2. Near-surface magmatism.
3. Mechanisms and timing of deformation.
4. Cause(s) of crustal magnetization.
5. Estimation of upper-crustal density.
6. Mass bounds on polar volatiles.

The science objective and six investigations are new to the GRAIL mission. The XM objectives and measurement requirements are shown in Table 5 and Fig. 11. These investigations do not cover the full scope of research that will be enabled by the GRAIL XM, but they are indicative of the kinds of analyses that will be possible by gravitational mapping of the Moon's upper crust with unprecedented resolution and accuracy.

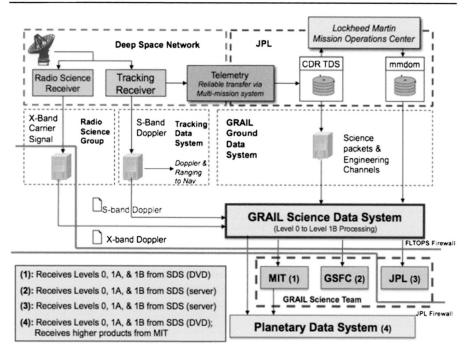

Fig. 12 GRAIL Science downlink data flow diagram

7 Data Flow and Processing

The GRAIL Science Data System (SDS) is the infrastructure at JPL for the collection of all science and ancillary data relevant to the GRAIL mission. The SDS includes hardware, software tools, procedures and trained personnel. The SDS receives data from three sources, collectively called Level 0 data, as described below, and carries out calibration, editing, and processing to produce Level-1A and -1B GRAIL science data. The SDS distributes Level-0, -1A, and -1B data to the Science Team and submits the same products for archiving at the NASA Planetary Data System (PDS) Geosciences Node. Higher-level data products including the gravitational field harmonic coefficients, are archived with the PDS by the Principal Investigator at MIT.

Figure 12 shows the downlinked data flow. The Mission Operation System (MOS) receives packets from the DSN and places them on the Telemetry Delivery System (TDS). The science data and engineering data packets are then transferred from the multimission TDS to the GRAIL science server. Timed scripts push the packets to the SDS computers on a regular basis. The SDS also receives Level-1A Doppler (tracking) data from the DSN. Finally, the SDS receives high-rate telemetry data from the Multimission Distributed Object Manager (MMDOM), placed there by the Lockheed Martin Mission Operations Center (MOC). Data are transferred to secure servers at MIT and the Goddard Space Flight Center for access and use by the GRAIL Science Team.

Table 6 GRAIL archive summary

Archive component	Data sets
LGRS (levels 0, 1A, & 1B)	Raw Ka-band phase, Time Transfer System (TTS) range data, and payload housekeeping information
	DSN Doppler data at S-band
DSN Tracking Data	DSN tracking data message files at X-band
	DSN Doppler data at X-band (optional)
	High-rate engineering data
	Engineering data
	Spacecraft properties
Ancillary	Spacecraft and planetary ephemeris
	DSN media calibration and Earth orientation parameters files
	Mission history log files
	Uplink products
	Data quality report files (levels 1A & 1B)
Software	None needed (files in ASCII format)
Documentation	GRAIL Gravity Theoretical Description and Data Processing Handbook, Software Interface Specifications (SIS), and calibration reports

8 Data Accessibility

The GRAIL mission will archive all acquired science data at all levels along with ancillary data and relevant information and documentation to NASA's Planetary Data System (PDS) in order for the science community at large to benefit from the knowledge gained by the mission. The archival will be complete and timely per NASA's guidelines and will allow future users to choose to either interpret available higher products or re-derive results from available lower data products. Additional documentation or software may also be provided at the discretion of the GRAIL Science Team.

The process of data accessibility in captured Table 6, which lists the archival data sets, and Table 7, which lists the data product identification for PDS labels, data levels, and expected volumes. Figure 12 shows the flow of data from the flight system to the ground system and ultimate users.

Images from the MoonKAM investigation are not required for the fulfillment of any GRAIL science objective and therefore are neither calibrated nor archived in the PDS. However, the images are posted as soon as possible after acquisition to a public website: http://images.moonkam.ucsd.edu/main.php, where they can be freely accessed by students and the public. Over 101,000 student images were acquired in the PM.

9 Summary

GRAIL successfully completed its Primary Mission on schedule and under budget. The mission achieved NASA's baseline mission success criteria (Investigation 1–4) for the PM in May 2012, one year ahead of schedule. GRAIL was successful in collecting its required data with a total science data volume at the end of the PM of 637 Mbytes or >99.99 % of possible data. Following the PM, the two GRAIL spacecraft successfully transited the

Table 7 GRAIL PDS data sets

Data set (volume ID) PDS-assigned data set ID	Description	Data volume (GB)	NASA processing level
LGRS EDR (GRAIL_0001) GRAIL-L-LGRS-2-EDR-V1.0	Raw science data in time order with duplicates and transmission errors removed	0.13	0
LGRS CDR (GRAIL_0101) GRAIL-L-LGRS-3-CDR-V1.0	Calibrated & resampled data	20	1A & 1B
RSS EDR (GRAIL_0201) GRAIL-L-RSS-2-EDR-V1.0	Raw Radio Science data (includes DSN Doppler tracking data, troposphere and ionosphere media calibrations)	137	0
LGRS SPICE (GRAIL_0301) GRAIL-L-SPICE-6-ADR-V1.0	SPICE geometry and navigation kernels (to be defined with help from NAIF)	6.7	N/A
LGRS RDR (GRAIL_1001) GRAIL-L-LGRS-5-RDR-GRAVITY-V1.0	Lunar gravitational field (includes gravity coefficient and covariance matrices, free-air gravity map, geoid and uncertainty maps, and Bouguer gravity map)	45	2

partial lunar eclipse of June 4, 2012 and initiated Extended Mission data gathering at very low mean altitude (23 km). Finally, GRAIL is on track to provide a comprehensive data set that will guide future scientific discoveries and future exploration of the Moon (Zuber et al. 2012).

Acknowledgements   The GRAIL mission is supported by NASA's Discovery Program and is performed under contract to the Massachusetts Institute of Technology. Part of this work was carried out at the Jet Propulsion Laboratory, California Institute of Technology, under a contract with the National Aeronautics and Space Administration.

Appendix

Acronyms and Abbreviations

CBE	Current Best Estimate
CCAFS	Cape Canaveral Air Force Station
C&DH	Command & Data Handling
CM	Center of Mass
DSN	Deep Space Network
ECM	Eccentricity Correction Maneuver
E/PO	Education and Public Outreach
GB	Gigabytes

GDS	Ground Data System
GPA	Gravity Processing Assembly
GR-A	GRAIL-A Spacecraft (Ebb)
GR-B	GRAIL-B Spacecraft (Flow)
GRACE	Gravity Recovery and Climate Experiment
GRAIL	Gravity Recovery and Interior Laboratory
GSFC	Goddard Space Flight Center
ITAR	International Traffic in Arms Regulations
JPL	Jet Propulsion Laboratory
KBR	Ka-Band Ranging
LGRS	Lunar Gravity Ranging System
LMSSC	Lockheed Martin Space Systems Company (Denver)
LOI	Lunar Orbit Insertion
LOLA	Lunar Orbiter Laser Altimeter
LOS	Line of Sight
LRO	Lunar Reconnaissance Orbiter
mascon	Mass Concentration
mGal	milliGal (where 1 Gal $= 0.01$ m s^{-2})
MIT	Massachusetts Institute of Technology
MoonKAM	Moon Knowledge Acquired by Middle school students
MOC	Mission Operations Center
MOS	Mission Operations System
MGSS	Multi-mission Ground System Services
MMDOM	Multimission Distributed Object Manager
MPST	Mission Planning and Sequence Team
MWA	Microwave Assembly
NAIF	Navigation and Ancillary Information Facility
NASA	National Aeronautics and Space Administration
OPR	Orbital Period Reduction
OTM	Orbit Trim Maneuver
PDS	Planetary Data System
PM	Primary Mission
RSB	Radio Science Beacon
SCT	Spacecraft Team
SDS	Science Data System
SIS	Software Interface Specification
SRS	Sally Ride Science
TCM	Trajectory Correction Maneuver
TDS	Telemetry Delivery System
TLC	Trans-Lunar Cruise
TSF	Transition to Science Formation
TSM	Transition to Science Maneuver
TTS	Time Transfer System
USO	Ultra-stable Oscillator
XM	Extended Mission
XSS-11	Experimental Satellite System 11

References

E.L. Akim, Determination of the gravitational field of the Moon from the motion of the artificial lunar satellite "Lunar-10". Dokl. Akad. Nauk SSSR **170** (1966)

M.P. Ananda, Lunar gravity: a mass point model. J. Geophys. Res. **82**, 3040–3064 (1977)

S.W. Asmar et al., The scientific measurement system of the Gravity Recovery and Interior Laboratory (GRAIL) mission. *Space Sci. Rev.* (2012, this issue)

M.J. Chung, S.J. Hatch, J.A. Kangas, S.M. Long, R.B. Roncoli, T.H. Sweetser, Trans-lunar cruise trajectory design of GRAIL mission, in *AIAA Astrodynamics Conf.*, Toronto, CA (2010)

C. Dunn et al., The instrument on NASA's GRACE mission: augmentation of GPS to achieve unprecedented gravity field measurements, in *Proc. 15th Int. Tech. Meeting of Satellite Division of Institute of Navigation*, Portland, OR, 2002, pp. 724–730

L.T. Elkins-Tanton, Linked magma ocean solidification and atmospheric growth for Earth and Mars. Earth Planet. Sci. Lett. **271**, 181–191 (2008)

GAO, *US Government Accountability Office Report to Congressional Committees, NASA—Assessments of Selected Large-Scale Projects* (U.S. Government Accountability Office, Washington, 2012), p. 16

S.J. Hatch, R.B. Roncoli, T.H. Sweetser, GRAIL trajectory design: lunar orbit insertion through science, in *AIAA Astrodynamics Conf.*, Toronto, CA (2010). AIAA 2010-8385, 8 pp.

W.A. Heiskanen, H. Moritz, *Physical Geodesy* (W.H. Freeman, San Francisco/London, 1967)

T.L. Hoffman, GRAIL: gravity mapping the Moon, in *IEEE Aerospace Conference*, Big Sky, MT, 978-1-4244-2622-5 (2009)

L.L. Hood, M.T. Zuber, Recent refinements in geophysical constraints on lunar origin and evolution, in *Origin of the Earth and Moon*, ed. by R.M. Canup, K. Righter (Univ. of Ariz. Press, Tucson, 2000), pp. 397–409

M.D. Johnson, J.E. Graf, R.W. Zurek, H.J. Eisen, B. Jai, The Mars Reconnaissance Orbiter mission, in *IEEE Aerospace Conf.* (2005)

W.M. Kaula, *Theory of Satellite Geodesy* (Blaisdell, Waltham, 1966), 124 pp.

W.M. Klipstein et al., The lunar gravity ranging system for the Gravity Recovery and Interior Laboratory (GRAIL) mission. *Space Sci. Rev.* (2012, this issue)

A.S. Konopliv, S.W. Asmar, E. Carranza, W.L. Sjogren, D.-N. Yuan, Recent gravity models as a result of the Lunar Prospector mission. Icarus **150**, 1–18 (2001)

A.S. Konopliv, A. Binder, L. Hood, A. Kucinskas, W.L. Sjogren, J.G. Williams, Gravity field of the Moon from Lunar Prospector. Science **281**, 1476–1480 (1998)

F.G. Lemoine, D.E. Smith, M.T. Zuber, G.A. Neumann, D.D. Rowlands, A 70th degree and order lunar gravity model from Clementine and historical data. J. Geophys. Res. **102**, 16339–16359 (1997)

E. Mazarico, F.G. Lemoine, S.-C. Han, D.E. Smith, GLGM-3, a degree-150 lunar gravity model from the historical tracking data of NASA Moon orbiters. J. Geophys. Res. **115**, E050001 (2010). doi:10.1029/2009JE003472

K. Matsumoto et al., An improved lunar gravity field model from SELENE and historical tracking data: revealing the farside gravity features. *J. Geophys. Res.* **115** (2010). doi:10.1029/2009JE003499

P.M. Muller, W.L. Sjogren, Mascons: lunar mass concentrations. Science **161**, 680–684 (1968)

N. Namiki et al., Farside gravity field of the Moon from four-way Doppler measurements of SELENE (Kaguya). Science **323**, 900–905 (2009)

G.A. Paulikas et al., *The Scientific Context for Exploration of the Moon: Final Report* (National Research Council, Washington, 2007), 120 pp.

R.J. Phillips, J.E. Conel, E.A. Abbot, W.L. Sjogren, J.B. Morton, Mascons: progress toward a unique solution for mass distribution. J. Geophys. Res. **77**, 7106–7114 (1972)

R.J. Phillips, W.L. Sjogren, E.A. Abbott, S.H. Zisk, Simulation gravity modeling to spacecraft tracking data: analysis and application. J. Geophys. Res. **83**, 5455–5464 (1978)

R.B. Roncoli, K.K. Fujii, Mission design overview for the Gravity Recovery and Interior Laboratory (GRAIL) mission, in *AIAA Guidance, Navigation and Control Conference, AIAA 2010-9393*, Toronto, Ontario, Canada (2010), 22 pp.

T.H. Sweetser, M.S. Wallace, S.J. Hatch, R.B. Roncoli, Design of an extended mission for GRAIL, in *AIAA Astrodynamics Specialist Conference, AIAA-2012-4439*, Minneapolis, MN (2012), 18 pp.

B.D. Tapley, S. Bettadpur, J.C. Ries, P.F. Thompson, M.M. Watkins, GRACE measurements of mass variability in the Earth system. *Science* **305** (2004). doi:10.1126/science.1099192

R.L. Taylor, M.T. Zuber, D.H. Lehman, T.L. Hoffman, Managing GRAIL: getting to launch on cost, on schedule and on spec, in *IEEE Aerospace Conference*, Big Sky, MT (2012)

J.B. Thomas, An analysis of gravity-field estimation based on intersatellite dual-1-way biased ranging, Jet Propulsion Laboratory (1999), 196 pp.

M.S. Wallace, T.H. Sweetser, R.B. Roncoli, Low lunar orbit design via graphical manipulation of eccentricity vector evolution, in *AIAA Astrodynamics Conference*, Minneapolis, MN (2012)

L. Wong, G. Buechler, W. Downs, W. Sjogren, P. Muller, P. Gottlieb, A surface layer representation of the lunar gravity field. J. Geophys. Res. **76**, 6220–6236 (1971)

J.A. Wood, J.S. Dickey, U.B. Marvin, B.N. Powell, Lunar anorthosites and a geophysical model for the Moon, in *Proc. Apollo 11 Lunar Sci. Conf.*, vol. 1 (1970), pp. 965–988

M.T. Zuber, D.E. Smith, F.G. Lemoine, G.A. Neumann, The shape and internal structure of the Moon from the Clementine mission. Science **266**, 1839–1843 (1994)

M.T. Zuber, D.E. Smith, D.H. Lehman, M.M. Watkins, Gravity Recovery and Interior Laboratory mission: facilitating future exploration to the Moon, in *Int. Astronaut. Congress*, Naples, Italy (2012)

DOI 10.1007/978-1-4614-9584-0_3
Reprinted from *Space Science Reviews* Journal, DOI 10.1007/s11214-013-9962-0

The Scientific Measurement System of the Gravity Recovery and Interior Laboratory (GRAIL) Mission

Sami W. Asmar · Alexander S. Konopliv · Michael M. Watkins · James G. Williams ·
Ryan S. Park · Gerhard Kruizinga · Meegyeong Paik · Dah-Ning Yuan ·
Eugene Fahnestock · Dmitry Strekalov · Nate Harvey · Wenwen Lu · Daniel Kahan ·
Kamal Oudrhiri · David E. Smith · Maria T. Zuber

Received: 5 October 2012 / Accepted: 22 January 2013 / Published online: 21 February 2013
© Springer Science+Business Media Dordrecht 2013

Abstract The Gravity Recovery and Interior Laboratory (GRAIL) mission to the Moon utilized an integrated scientific measurement system comprised of flight, ground, mission, and
data system elements in order to meet the end-to-end performance required to achieve its
scientific objectives. Modeling and simulation efforts were carried out early in the mission
that influenced and optimized the design, implementation, and testing of these elements.
Because the two prime scientific observables, range between the two spacecraft and range
rates between each spacecraft and ground stations, can be affected by the performance of
any element of the mission, we treated every element as part of an extended science instrument, a science system. All simulations and modeling took into account the design and
configuration of each element to compute the expected performance and error budgets. In
the process, scientific requirements were converted to engineering specifications that became the primary drivers for development and testing. Extensive simulations demonstrated
that the scientific objectives could in most cases be met with significant margin. Errors are
grouped into dynamic or kinematic sources and the largest source of non-gravitational error comes from spacecraft thermal radiation. With all error models included, the baseline
solution shows that estimation of the lunar gravity field is robust against both dynamic and
kinematic errors and a nominal field of degree 300 or better could be achieved according to
the scaled Kaula rule for the Moon. The core signature is more sensitive to modeling errors
and can be recovered with a small margin.

Keywords Gravity · Moon · Remote sensing · Spacecraft

Acronyms and Abbreviations
AGC   Automatic Gain Control

S.W. Asmar (✉) · A.S. Konopliv · M.M. Watkins · J.G. Williams · R.S. Park · G. Kruizinga · M. Paik ·
D.-N. Yuan · E. Fahnestock · D. Strekalov · N. Harvey · W. Lu · D. Kahan · K. Oudrhiri
Jet Propulsion Laboratory, California Institute of Technology, Pasadena, CA 91109, USA
e-mail: sami.asmar@jpl.nasa.gov

D.E. Smith · M.T. Zuber
Department of Earth, Atmospheric and Planetary Sciences, Massachusetts Institute of Technology,
Cambridge, MA 02139-4307, USA

AMD	Angular Momentum Desaturation
C&DH	Command & Data Handler
CBE	Current Best Estimate
CG	Center of Gravity
CPU	Central Processing Unit
DSN	Deep Space Network
DOWR	Dual One-way Range
ECM	Eccentricity Correction Maneuver
EOP	Earth Orientation Platform
ET	Ephemeris Time
GPS	Global Positioning System
GR-A	GRAIL-A Spacecraft (Ebb)
GR-B	GRAIL-B Spacecraft (Flow)
GRACE	Gravity Recovery and Climate Experiment
GRAIL	Gravity Recovery and Interior Laboratory
GSFC	Goddard Space Flight Center
ICRF	International Celestial Reference Frame
IERS	International Earth Rotation and Reference Systems Service
IR	Infra Red
IPU	Instrument Processing Unit (GRACE mission)
JPL	Jet Propulsion Laboratory
KBR	Ka-Band Ranging
KBRR	Ka-Band Range-Rate
LGRS	Lunar Gravity Ranging System
LLR	Lunar Laser Ranging
LOI	Lunar Orbit Insertion
LOS	Line of Sight
LP	Lunar Prospector
mGal	milliGal (where $1 \text{ Gal} = 0.01 \text{ m s}^{-2}$)
MGS	Mars Global Surveyor
MIT	Massachusetts Institute of Technology
MOS	Mission Operations System
MIRAGE	Multiple Interferometric Ranging and GPS Ensemble
MMDOM	Multi-mission Distributed Object Manager
MONTE	Mission-analysis, Operations, and Navigation Toolkit Environment
MPST	Mission Planning and Sequence Team
MRO	Mars Reconnaissance Orbiter
MWA	Microwave Assembly
NASA	National Aeronautics and Space Administration
ODP	Orbit Determination Program
OPR	Orbital Period Reduction
OSC	Onboard Spacecraft Clocks
OTM	Orbit Trim Maneuver
PDS	Planetary Data System
PM	Primary Mission
PPS	Pulse Per Second
RSB	Radio Science Beacon
RSR	Radio Science Receiver
SCT	Spacecraft Team

SDS	Science Data System
SIS	Software Interface Specification
SRIF	Square Root Information Filter
SRP	Solar Radiation Pressure
TAI	International Atomic Time
TCM	Trajectory Correction Maneuver
TDB	Barycentric Dynamic Time
TDS	Telemetry Delivery System
TDT	Terrestrial Dynamic Time
TLC	Trans-Lunar Cruise
TSF	Transition to Science Formation
TSM	Transition to Science Maneuver
TTS	Time Transfer System
USO	Ultra-stable Oscillator
UTC	Universal Time Coordinated
VLBI	Very Long Baseline Interferometry

1 Introduction and Heritage

The Gravity Recovery and Interior Laboratory (GRAIL) mission is comprised of two spacecraft, named Ebb and Flow, flying in precision formation around the Moon. The mission's purpose is to recover the lunar gravitational field in order to investigate the interior structure of the Moon from the crust to the core. The spacecraft were launched together on September 10, 2011 and began science operations and data acquisition on March 1, 2012. Zuber et al. (2013, this issue) presents an overview of the mission including scientific objectives and measurement requirements. Klipstein et al. (2013, this issue) describes the design and implementation of the GRAIL payload. Hoffman (2009) described GRAIL's flight system and Roncoli and Fujii (2010) described the mission design.

This paper illustrates how a team of scientists and engineers prepared to meet GRAIL scientific objectives and data quality requirements through simulations and modeling of the design and configuration of the flight and ground systems. It details dynamic and kinematic models for estimating error sources in the form of non-gravitational forces and how these models were applied, along with the lunar gravity model, to elaborate computer simulations in the context of an integrated scientific measurement system. This paper also documents the methods, tools, and results of the simulations. This work was carried out at the Jet Propulsion Laboratory (JPL) prior to the science orbital phase and reviewed by expert peers from different institutions; the knowledge is based on the combined experiences of the team members with gravity observations on numerous planetary missions. This effort demonstrated that the mission was capable of meeting the science requirements as well as paved the way to the operational tools and procedures for the actual science data analysis.

The GRAIL concept was derived from the Gravity Recovery and Climate Experiment (GRACE) Earth mission and utilized a modified GRACE payload called the Lunar Gravity Ranging System (LGRS); the GRAIL and GRACE spacecraft are unrelated. For an overview of the GRACE mission see Tapley et al. (2004a, 2004b); for a description of the GRACE payload, see Dunn et al. (2003); and for error analysis in the GRACE system and measurements, see Kim and Tapley (2002).

Despite the high heritage, there are significant differences between the GRAIL and GRACE science payloads, listed in Table 1. GRACE is equipped with a Global Positioning

Table 1 Functional differences between the GRAIL and GRACE Missions

	GRAIL	GRACE
Target body	Moon	Earth
Launch vehicle	Delta II, USA	Rockot, Russia
Nominal prime mission duration	3 months	5 years
Orbiter mass (kg)	313	487
Launch date	9/10/11	3/17/02
Prime mission mean orbital altitude (km)	55	470
Gravity coefficients	420	120
Timing synchronization method	RSB	GPS
Science-quality accelerometer	N	Y
Adjustable mass for accelerometer at CG	N	Y
Center of gravity calibrations for antenna	Y	Y
Inter-spacecraft links	Ka-/S-band	Ka-/K-band
Spacecraft separation distance (km)	85–225	170–270
Attitude control	Reaction wheels	Magnetic torque
Thrusters gas	Hydrazine	Nitrogen
Star cameras per spacecraft	1	2
Science processor	Single String	Redundant
Star camera software host	C&DH[a]	IPU[b]
USOs per spacecraft	1	2
Absolute timing accuracy	DSN: millisecond	GPS: nanosecond
Relative timing accuracy	TTS: picosecond	GPS: picosecond
Communication stations	DSN	German stations

[a]C&DH is GRAIL's Command and Data Handling Subsystem.

[b]IPU is GRACE's Instrument Processing Unit.

System (GPS) receiver for timing synchronization, and accelerometers for non-gravitational force calibrations, while GRAIL is not. Furthermore, GRACE inter-spacecraft ranging utilizes two radio links at K- and Ka-bands ($\sim$26 GHz and $\sim$32 GHz, respectively) in order to calibrate the effects of charged particles in the Earth ionosphere, while GRAIL utilizes only one Ka-band link. In lieu of GPS time synchronization, which is not available at the Moon, GRAIL introduced two elements, a second inter-spacecraft link at S-band ($\sim$2.3 GHz) for a Time Transfer System (TTS), and a one-way X-band ($\sim$8.4 GHz) link transmitted from each spacecraft's Radio Science Beacon (RSB) to the Deep Space Network (DSN) stations. With these differences, the GRAIL observable time tagging and synchronization is handled differently from the GRACE GPS-based system as will be discussed below.

Furthermore, while the GRACE observables are referenced to a geocentric frame, GRAIL measurements are referenced to Ephemeris Time (ET) and the solar system barycentric frame of Barycentric Dynamic Time (TDB). Finally, since GRAIL does not carry an accelerometer, attention was given in the design, assembly, and testing of the spacecraft system in order to minimize on the non-gravitational forces acting on the spacecraft, including the solar radiation pressure, lunar albedo and spacecraft outgassing.

All radio signals in the science payload, illustrated in Fig. 1, the Ka-band inter-spacecraft link, the S-band TTS inter-spacecraft link, and the X-band RSB link to Earth, are referenced

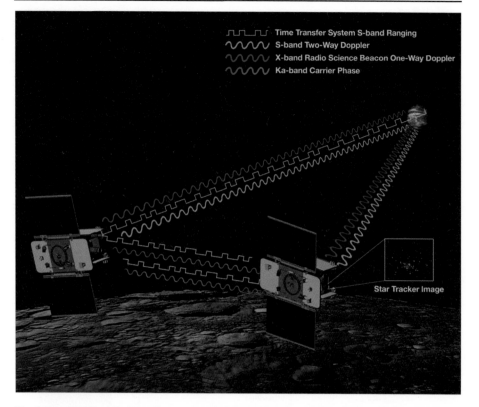

Fig. 1 The GRAIL radio links: Ka-band and S-band inter-spacecraft links, X-band one-way downlink to ground stations, and two-way S-band links for telecommunications and navigation

on one Ultra-Stable Oscillator (USO) per spacecraft. The navigation and telecommunications telemetry and command functions are handled by a separate two-way S-band link with the DSN. This is a spacecraft system function not linked to the science payload or the USO. The science data quality simulations did not incorporate the utility of this telecommunications link since no science performance requirements were imposed on it, but in reality, the science team collaborated with the navigation team to assess its usability to enhance the science results.

2 Simulations Tool and Data Levels

Over several decades, NASA's JPL has developed techniques, algorithms, and software tools to conduct investigations of planetary gravitational fields and applied them to practically every planet in the solar system and several satellites of the outer planets. JPL relies primarily on the Orbit Determination Program (ODP) whose formulation is detailed in Moyer (2003). The ODP has enabled precision navigation for the vast majority of deep space missions and, due to its criticality to the success of these missions, has received rigorous development and testing as well as continued improvements (a new tool called Mission-analysis, Operations, and Navigation Toolkit Environment, or MONTE, has replaced the ODP for mission navigation purposes).

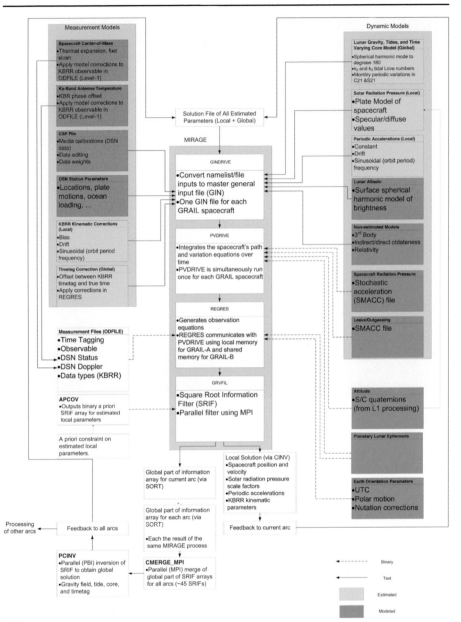

Fig. 2 A functional flowchart of the MIRAGE software tool as used in the simulation process

The GRAIL scientists at JPL use a version of the ODP called Multiple Interferometric Ranging and GPS Ensemble (MIRAGE), which originated from a GPS version of the ODP developed for the TOPEX mission (described in Guinn and Wolff 1993, and Leavitt and Salama 1993) and further developed for gravity field analysis, (Fahenstock 2009). Figure 2 shows the MIRAGE flowchart process utilized for GRAIL and the various programs that process the generalized inputs, the spacecraft path integration, computation of dynamic parameter partials, and the data observables. This figure documents the necessary interfaces between the software elements and input/output files as well as the relevant computational parameters and has been a key figure for the simulations peer-review process. There are three subsets of programs that integrate the spacecraft motion, process the spacecraft observations and filter or estimate the spacecraft state and related parameters using the observations.

To determine the spacecraft dynamical path, the program numerically integrates the spacecraft Cartesian state by including all known forces acting on the spacecraft, such as gravity, solar pressure, lunar albedo, and spacecraft thrusting. The spacecraft state and the force model partial derivatives (e.g., gravity harmonics) that are later estimated are integrated using the variable order Adams method described in Krogh (1973). The non-rotating International Celestial Reference Frame (ICRF) defines the inertial coordinate system, which is nearly equal to the Earth's mean equator and equinox at the epoch of J2000.

The GRAIL data are categorized in 3 levels, also shown in of Zuber et al. (2013, this issue). Level 0 is the raw data acquired by the spacecraft science payload, the LGRS, and DSN Doppler. Level 1 is the expanded, edited and calibrated data. Level 1 processing is the conversion from Level 0 files to Level 1 files. Level 1 processing also applies a time tag conversion, time of flight correction, and phase center offset, as well as generates instantaneous range-rate and range-acceleration observables by numerical differentiation of the biased range observables. Level 2 is the gravity field spherical harmonic expansion; level 2 processing refers to the production of Level 2 data. The simulations described herein emulate the generation of Levels 1 and 2 GRAIL mission data.

3 Gravity Model Representation

Gravitational fields provide a key tool for probing the interior structure of planets. The lunar gravity, when combined with topography, leads to geophysical models that address important phenomena such as the structure of the crust and lithosphere, the asymmetric lunar thermal evolution, subsurface structure of impact basins and the origin of mascons, and the temporal evolution of crustal brecciation and magmatism. Long-wavelength gravity measurements can place constraints on the presence of a lunar core.

A gravitational field represents variations in the gravitational potential of a planet and gravity anomalies at its surface. It can be mathematically represented via coefficients of a spherical harmonic expansion whose degree and order reflect the surface resolution. A field of degree 180, for example, represents a half-wavelength, or spatial block size, surface resolution of 30 km; for degree n, the resolution is $30 \times 180/n$ km. The gravitational potential in spherical harmonic form is represented in the body-fixed reference frame with normalized coefficients $(\overline{C}_{nm}, \overline{S}_{nm})$ is represented after Heiskanen and Moritz (1967) and Kaula (1966) as:

$$U = \frac{GM}{r} + \frac{GM}{r} \sum_{n=1}^{\infty} \sum_{m=0}^{n} \left(\frac{R_e}{r} \right)^n \overline{P}_{nm}(\sin \varphi_{lat}) \left[\overline{C}_{nm} \cos(m\lambda) + \overline{S}_{nm} \sin(m\lambda) \right] \quad (1)$$

G is the gravitational constant, M is the mass of the central body, r is the radial distance coordinate, m is the order, $\overline{P}_{nm}$ are the fully normalized associated Legendre polynomials, R_e is the reference radius of the body, φ_{lat} is the latitude, and λ is the longitude. The gravity coefficients are normalized so that the integral of the harmonic squared equals the area of a unit sphere, and are related to the un-normalized coefficients by Kaula (1966), where δ is the Kronecker delta:

$$\begin{pmatrix} C_{nm} \\ S_{nm} \end{pmatrix} = \left[\frac{(n-m)!(2n+1)(2-\delta_{0m})}{(n+m)!} \right]^{1/2} \begin{pmatrix} \overline{C}_{nm} \\ \overline{S}_{nm} \end{pmatrix} = f_{nm} \begin{pmatrix} \overline{C}_{nm} \\ \overline{S}_{nm} \end{pmatrix} \tag{2}$$

There exist singularities at the pole in the partials of the gravity acceleration with respect to the spacecraft position when using the Legendre polynomials as a function of latitude. To accommodate this, MIRAGE uses a nonsingular formulation of the gravitational potential, including recursion relations given by Pines (1973), in calculation of the acceleration and partials.

The gravitational potential also accounts for tides caused by a perturbing body. The second-degree tidal potential acting on a satellite at position $\vec{r}$ relative to the central body, with the perturbing body (e.g., Sun and Earth for GRAIL) at position $\vec{r}_p$, is:

$$U = k_2 \frac{GM_p}{R} \frac{R^6}{r^3 r_p^3} \left[\frac{3}{2} (\hat{r} \cdot \hat{r}_p)^2 - \frac{1}{2} \right] \tag{3}$$

where k_2 is the second degree potential Love number, M_p is the mass of the perturbing body causing the tide, and R is the equatorial radius of the central body. Tides raised on the Moon by the Sun are two orders-of-magnitude smaller than tides raised by the Earth. The acceleration due to constant lunar tides is modeled using a spherical harmonics representation:

$$\Delta C_{nm} - i \Delta S_{nm} = \frac{k_{nm}}{2n+1} \sum_j \frac{GM_j}{GM} \frac{R_M^{n+1}}{r_{mj}^{n+1}} P_{nm}(\sin \varphi_j) e^{-im\lambda_j} \tag{4}$$

Simplifying, the non-dissipative tides contribute time-varying components to second degree and order normalized coefficients as follows (McCarthy and Petit 2003):

$$\Delta \overline{J}_2 = -k_{20} \sqrt{\frac{1}{5}} \frac{GM_p R^3}{GM r_p^3} \left[\frac{3}{2} \sin^2 \varphi_p - \frac{1}{2} \right]$$

$$\Delta \overline{C}_{21} = k_{21} \sqrt{\frac{3}{5}} \frac{GM_p R^3}{GM r_p^3} \sin \varphi_p \cos \varphi_p \cos \lambda_p$$

$$\Delta \overline{S}_{21} = k_{21} \sqrt{\frac{3}{5}} \frac{GM_p R^3}{GM r_p^3} \sin \varphi_p \cos \varphi_p \sin \lambda_p \tag{5}$$

$$\Delta \overline{C}_{22} = k_{22} \sqrt{\frac{3}{20}} \frac{GM_p R^3}{GM r_p^3} \cos^2 \varphi_p \cos 2\lambda_p$$

$$\Delta \overline{S}_{22} = k_{22} \sqrt{\frac{3}{30}} \frac{GM_p R^3}{GM r_p^3} \cos^2 \varphi_p \sin 2\lambda_p$$

Here, φ_p and λ_p are the latitude and longitude of the perturbing body on the surface of the central body. Separate Love numbers have been used for each order, though they are expected to be equal ($k_{20} = k_{21} = k_{22}$). Degree-3 Love number solutions have been investigated and their effect is barely detectable.

The tidal potential consists of a variable term and a constant or permanent term. Depending on choice of convention, the constant term may or may not be included in the corresponding gravity coefficient. The MIRAGE-generated gravity fields do not include the

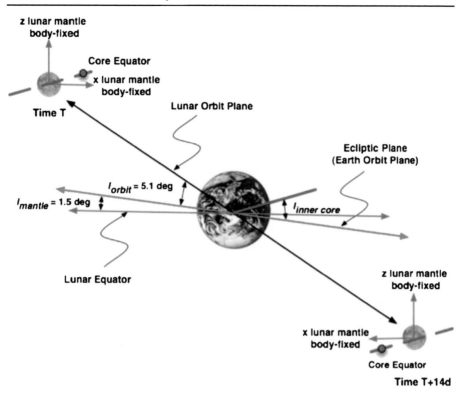

Fig. 3 Possible lunar core motion and the relationship between different frames of reference

permanent part of the tide. Our formulation assumes an elastic Moon and does not include the frequency-dependent dissipation terms. The elasticity does not affect the overall simulation results and was not considered in this study. The k_2 estimate uncertainty from lunar laser ranging and spacecraft tracking is between 6–8 percent. The GRAIL results will determine the k_2 Love number to better than 1 percent.

The acceleration due to the gravitational potential must be rotated from the body-fixed principal axis frame to the inertial frame using the lunar physical libration angles included in a planetary ephemeris database (e.g., JPL DE421) described in Williams et al. (2008). Three Euler angles describe lunar orientation: the angle along the J2000 equator from the J2000 equinox to the intersection of the lunar equator with the J2000 equator, the angle between the two equators, and the angle along the lunar equator from the intersection of equators to the lunar meridian of zero longitude (Newhall and Williams 1997).

On the basis of re-analysis of Apollo seismic observations, Weber et al. (2011) proposed that the Moon has a solid inner core surrounded by a fluid outer core. Given an oblate inner core, a time-varying signature could result from the monthly motion of the lunar core equator relative to the lunar body-fixed or mantle frame (Williams 2007), affecting the degree 2 and order 1 spherical harmonics and the second-degree tidal potential changes due to the Earth and Sun. Figure 3 illustrates the Moon's expected core motion; a point on the core equator moves relative to the body-fixed equator with a period of one month.

Due to the pole offset of the core and mantle frame, the core motion introduces a monthly signature in the $\overline{C}_{21}$ and $\overline{S}_{21}$ gravity coefficients as follows:

$$\Delta\overline{C}_{21} = \alpha_{21}\cos(\dot{\omega}t + \varphi), \tag{6}$$

$$\Delta\overline{S}_{21} = \beta_{21}\cos(\dot{\omega}t + \varphi) \tag{7}$$

where $\Delta\overline{C}_{21}, \Delta\overline{S}_{21}$ is the monthly gravitational potential oscillation due to a possible solid inner core with an axis of rotation tilted relative to the mantle's axis, included in all simulations, $\dot{\omega}$ is the frequency and φ is the phase of this periodic signature. For the latter, we assume a priori knowledge when estimating the amplitudes of the $\overline{C}_{21}$ and $\overline{S}_{21}$ signatures (α_{21} and β_{21}) along with the gravity field and tidal Love number. If the inner core had an equilibrium figure for tide and spin distortion, then the ratio of amplitudes for $\overline{C}_{21}$ and $\overline{S}_{21}$ signatures would be 4. While this ratio is not assumed, it has been used to set requirements for amplitude uncertainties. We investigated both the uncertainty of the core amplitudes and the differences of the estimated values with the a priori values. These estimated amplitudes plus the tidal Love numbers encapsulate the results of GRAIL's science investigations addressing the deep interior.

4 Model Estimation and Dynamical Integration

JPL's gravity field estimation process relies on two primary data types: a link between the spacecraft and Earth, which is a one-way X-band link, and an inter-spacecraft link called the Ka-band Range (KBR). The latter's first derivative, the Ka-band Range Rate (KBRR), precisely measures the relative movement of Ebb and Flow, which permits estimation of the lunar gravity field. The combined measurement of two sets of ranging data, one measured by Ebb and a second by Flow, is called the Dual One-Way Range (DOWR) measurement. Ebb and Flow are tracked from Earth by the DSN, which produces Doppler data used to determine the absolute position of each spacecraft:

$$z_d = \hat{\rho}_{se} \cdot \dot{\rho}_{se}, \tag{8}$$

$$z_s = \hat{\rho}_{ba} \cdot \dot{\rho}_{ba} \tag{9}$$

where ρ_{se} represents the vector from spacecraft to the DSN station and ρ_{ba} represents the vector from Ebb to Flow.

The estimation of the gravity field follows the same steps as the orbit determination process in navigation but involves many more parameters and methods that may constrain the gravity field and other model parameters to obtain the most realistic solution. Although the planetary gravity field solutions often require a Kaula power law constraint (Kaula 1966), the uniform and global coverage of the KBRR data does not require a constraint in our simulations except for solutions of high degree (i.e., degree $\sim$270) where a small power-type constraint was applied.

Letting $\vec{r}$ and $\vec{v}$ be the position and velocity vectors of the spacecraft relative to the central body, the software integrates the second order differential equations

$$\ddot{\vec{r}} = \vec{f}(\vec{r}, \vec{v}, \vec{q}) = \nabla U(\vec{r}) + \vec{f}_{pm} + \vec{f}_{in\text{-}pm} + \vec{f}_{in\text{-}obl} + \vec{f}_{srp} + \vec{f}_{alb} + \vec{f}_{att} + \vec{f}_{rel} + \cdots \tag{10}$$

Here, $\vec{f}(\vec{r}, \vec{v}, \vec{q})$ is the total acceleration of the spacecraft and $\vec{q}$ are all the constant ($\dot{\vec{q}} = 0$) model parameters to be estimated (e.g., gravity harmonic coefficients). Contributions to the total acceleration include the acceleration of the spacecraft relative to the central body due to the gravitational potential of the central body $\nabla U(\vec{r})$, the spacecraft acceleration due to other solar system bodies treated as point masses $\vec{f}_{pm}$, the indirect point mass acceleration of the central body in the solar system barycentric frame due to the other planets and natural satellites $\vec{f}_{in\text{-}pm}$, the indirect oblateness acceleration of the central body (e.g., Moon) due to another body's oblateness (e.g., Earth) $\vec{f}_{in\text{-}obl}$, the acceleration of the spacecraft due to solar radiation pressure $\vec{f}_{srp}$, the acceleration due to lunar albedo $\vec{f}_{alb}$, the acceleration due to spacecraft gas thrusting for attitude control maneuvers (usually for de-spinning angular momentum wheels) $\vec{f}_{att}$, and the pseudo-acceleration due to general relativity corrections $\vec{f}_{rel}$. Other accelerations also exist and may include spacecraft thermal forces, infrared radiation, tides, and empirical, usually periodic, acceleration models. Specific acceleration models that have been taken into account are described below.

4.1 Acceleration Due to Solar Radiation Pressure

Each spacecraft is modeled with five single-sided flat plates to model the acceleration due to solar radiation pressure (SRP) as detailed in Fahnestock et al. (2012) and Park et al. (2012). For each plate, the acceleration is computed as:

$$a_{srp} = \frac{C S_s}{m_s r_{sp}^2} (F_n \hat{u}_n + F_r \hat{u}_s), \tag{11}$$

$$F_n = -A(2\kappa_d v_d + 4\kappa_s v_s \cos\alpha)\cos\alpha, \tag{12}$$

$$F_r = -A(1 - 2\kappa_s v_s)\cos\alpha. \tag{13}$$

The acceleration due to SRP is on the order of 10^{-10} km/s^2. It is separable from the effect of gravity in the estimation process. With a ray-tracing technique to model self-shadowing on the spacecraft bus and on-board telemetry of the power system to detect entry and exit from lunar shadow, the SRP accelerations can be determined to a few percent level.

4.2 Acceleration Due to Spacecraft Thermal Radiation:

For a flat plate component, the acceleration due to spacecraft thermal re-radiation is:

$$a_{str} = \frac{-2 \times 10^{-6} A \sigma_{sb}}{3 m_s c} \varepsilon T^4 \hat{u}_n. \tag{14}$$

This is used to convert from any given plate's surface temperature to its acceleration contribution.

4.3 Acceleration Due to Lunar Albedo and Thermal Emission

The element of acceleration on a spacecraft due to lunar radiation pressure from a point P on the surface of the Moon can be computed (from Park et al. 2012) as:

$$da_{lrp} = H(F_n \hat{u}_n + F_r \hat{r}_{ps}) \frac{\cos\psi}{\pi r_{ps}^2} dA_{planet}. \tag{15}$$

For reflected sunlight (albedo):

$$H = \frac{C S_m \cos\psi_s}{m_s r_{ms}^2} \sum_{\ell=0}^{N} \sum_{m=0}^{\ell} (C_{\ell m}^A \cos m\lambda_p + S_{\ell m}^A \sin m\lambda_p) P_{\ell m}(\sin\varphi_p), \tag{16}$$

and for thermal emission (infrared):

$$H = \frac{C}{4m_s r_{ms}^2} \sum_{\ell=0}^{N} \sum_{m=0}^{\ell} \left(C_{\ell m}^E \cos m\lambda_p + S_{\ell m}^E \sin m\lambda_p \right) P_{\ell m}(\sin \varphi_p). \tag{17}$$

The albedo map is a constant field whereas the thermal map is a function of local lunar time because of topographic variation; the thermal map derived using the measurements from Lunar Reconnaissance Orbiter's Diviner Lunar Radiometer Experiment data. For this reason, the following simplified thermal emission model was derived for the simulation of the total error budget:

$$H = \begin{cases} \frac{C\sigma_{sb} T_{max}^4 \cos \psi_s}{4m_s r_{cs}^2 L}, & \text{if } \psi_s \leq 89.5°, \\ \frac{C\sigma_{sb} T_{min}^4}{4m_s r_{cs}^2 L}, & \text{otherwise,} \end{cases} \tag{18}$$

where $T_{max} = 382.86$ K and $T_{min} = 95$ K. Thermal maps were computed at the local noon-time when the Sun is at 0° longitude and 0° latitude.

4.4 Acceleration Due to Un-modeled Forces

The acceleration due to un-modeled forces is used to represent the errors in the non-gravitational forces from solar pressure, spacecraft thermal radiation, lunar radiation, and spacecraft outgassing and is represented as the periodic acceleration formulation:

$$\begin{aligned} a_{uf} = & (P_r + C_{r1} \cos \theta + C_{r2} \cos 2\theta + S_{r1} \sin \theta + S_{r2} \sin 2\theta)\hat{e}_r \\ & + (P_t + C_{t1} \cos \theta + C_{t2} \cos 2\theta + S_{t1} \sin \theta + S_{t2} \sin 2\theta)\hat{e}_t \\ & + (P_n + C_{n1} \cos \theta + C_{n2} \cos 2\theta + S_{n1} \sin \theta + S_{n2} \sin 2\theta)\hat{e}_n, \end{aligned} \tag{19}$$

where $\hat{e}_r$, $\hat{e}_t$, and $\hat{e}_n$ represent the radial, transverse, and normal unit-vectors, respectively and θ denotes the angle from the ascending node of the spacecraft orbit on the EME2000 plane to the spacecraft. The periodic acceleration is nominally set to zero in the initial trajectory integration and is used to estimate the errors in the non-gravitational accelerations. The terms P_i represent the constant accelerations during the time interval that the corresponding periodic acceleration model is active. The terms (C_{i1}, S_{i1}) and (C_{i2}, S_{i2}) represent the once-per-orbit and twice-per-orbit acceleration amplitudes, respectively.

In addition to integrating the spacecraft position and velocity, MIRAGE integrates the variational equations to estimate the epoch state and constant parameters. Following nomenclature in Tapley et al. (2004a, 2004b), the nominal trajectory is given by:

$$X^*(t) = \begin{pmatrix} \vec{r}^*(t) \\ \vec{v}^*(t) \\ \vec{q}^* \end{pmatrix}. \tag{20}$$

The first order differential equation to integrate in order to determine the nominal orbit is:

$$\dot{X}^*(t) = \begin{pmatrix} \vec{v}^* \\ \vec{f}(\vec{r}^*, \vec{v}^*, \vec{q}^*) \\ 0 \end{pmatrix} = F(X^*, t). \tag{21}$$

The variation of the trajectory from its nominal path is $x(t) = X(t) - X^*(t)$ and the linearized equations:

$$\dot{x}(t) = A(t)x(t) = \left(\frac{\partial F(t)}{\partial X(t)}\right)^* x(t) \tag{22}$$

The integrated solution is the state transition matrix $\Phi(t, t_0)$, which relates the deviation from the nominal path at epoch t_0 to the deviation from the nominal path at time t for the 6 position and velocity epoch parameters matrix $(U_{6\times6})$ and the p constant model parameters $(V_{6\times6})$:

$$x(t) = \Phi(t, t_0)x(t_0) = \begin{bmatrix} U_{6\times6} & V_{6\times6} \\ 0_{p\times6} & I_{p\times p} \end{bmatrix} x(t_0). \tag{23}$$

The second order differential equations that MIRAGE integrates for each GRAIL spacecraft include the 3 position variables of Eq. (10), 18 variables representing the changes in position and velocity due to small changes in epoch position and velocity which define the matrix $U_{6\times6}$, and three equations for each dynamic parameter or constant from being estimated. For a complete gravity field of degree and order n, the total number of gravity field parameters is given by $(n-1)(n+3)$, or, for example, 32,757 parameters for a 180 degree and order field.

5 Processing and Filtering of Observations

After numerical integration, MIRAGE processes Doppler and range observations. Following Tapley et al. (2004a), the general form of the observation equation is

$$Y = G(X, t) + \varepsilon, \tag{24}$$

where Y is the actual observation, $G(X, t)$ is a mathematical expression to calculate the modeled observation value, and ε is the observation error. The DSN Doppler data is not an instantaneous velocity measurement, but is processed in similar fashion to a range observable and is given by a differenced range measurement for two-way Doppler as

$$G(X, t) = \left((r_{12} + r_{23})_e - (r_{12} + r_{23})_s\right)/\Delta t + \cdots \tag{25}$$

where r_{12} is the uplink range transmitted by the ground station and received at the spacecraft, and r_{23} is the downlink range from the spacecraft to the earth station, with subscripts denoting the end and start of the Doppler count interval, Δt. To process a Doppler observation, we must solve the light time equation in a solar system barycentric frame, i.e., find the original transmit time at the first station and the receive time at the spacecraft using an iterative procedure. Equation (25) requires DSN calibrations for Earth ionospheric and tropospheric refraction (Mannucci et al. 1998), and corrections for relativistic propagation delay due to the Sun and planets, solar plasma delays due to the solar corona of the Sun, and any measurement biases.

The dual one-way phase measurement between Ebb and Flow can be converted to a biased range, by an algorithm first developed by Kim (2000). Our lunar gravity recovery process ingests instantaneous range-rate, modeled as a projection of the velocity difference vector, $\dot{r}_{12}$, along the line-of-sight unit vector, $\hat{e}_{12}$.

$$G(X, t) = \dot{\rho} = \dot{r}_{12} \bullet \hat{e}_{12} \tag{26}$$

Processing observables also requires the linearized form of Eq. (24). Given an observable Y, we compute a nominal observable $Y^*(t)$ based on an input nominal orbit, and calculate an observation residual y:

$$y = Y - Y^*(t) \tag{27}$$

Using the state transition matrix to map to the epoch time, Eq. (24) is then written as

$$y = \left(\frac{\partial G}{\partial X}\right)\Phi(t, t_0)x_0 + \varepsilon = Hx_0 + \varepsilon. \tag{28}$$

Based on the vector of residuals y and partials matrix H, the MIRAGE filter solves for a state X that minimizes these ε error terms.

The calculation of the nominal DSN Doppler observable and related partials in Eq. (28) involves the precise location of the Earth station in a solar system barycentric ICRF frame as shown in Yuan et al. (2001). The Earth-fixed coordinate system is consistent with the International Earth Rotation and Reference Systems Service (IERS) terrestrial reference frame labeled ITRF93 as shown in Boucher et al. (1994). The rotation of the Earth-fixed coordinates of the DSN locations to the Earth centered inertial system requires a series of coordinate transformations due to precession as in the IAU 1976 model described in Lieske et al. (1977) and nutation of the mean pole as in the IAU 1980 nutation theory described in Wahr (1981) and Seidelmann (1982) plus daily corrections to the model from the JPL Earth Orientation Platform (EOP) product of Folkner et al. (1993), rotation of the Earth as in Aoki et al. (1982) and Aoki and Kinoshita (1983) and UTC-UT1R corrections of the JPL EOP file, and polar motion of the rotation axis. The JPL EOP product is derived from the Very Long Baseline Interferometry (VLBI) and Lunar Laser Ranging (LLR) observations and includes Earth rotation and polar motion calibrations and, in addition, nutation correction parameters necessary to determine inertial station locations to the level of a few centimeters.

The body-fixed ITRF93 DSN station locations have been determined with VLBI measurements and conventional and GPS surveying. The coordinate uncertainties are about 4 cm for DSN stations that have participated in regular VLBI experiments, and about 10 cm for other stations; Folkner (1996) also provides the antenna phase center offset vector for each DSN station. These DSN station locations are consistent with the NNR-NEWVAL1 plate motion model (Argus and Gordon 1991). The variations of DSN station coordinates caused by solid Earth tide, ocean tide loading, and rotational deformation due to polar motion are corrected according to the IERS standards for 1992 (McCarthy and Petit 2003).

Once the observation equations are found, MIRAGE estimates the spacecraft state and other parameters using a weighted Square Root Information Filter (SRIF), see Lawson and Hanson (1995). SRIF computation time dominates MIRAGE processing, and for the larger planetary gravity fields of the Moon we run on two Beowulf Linux clusters (a 28-node machine with 112 CPU cores and a 45-node machine with 360 CPU cores). In normal form, the least-squares solution is given by:

$$\hat{x} = \left(H^T W H + P_{ap}^{-1}\right)^{-1} H^T W y \tag{29}$$

W is the weight matrix for the observations and P_{ap} is the a priori covariance matrix of the parameters being estimated. In the MIRAGE SRIF filter, the solution equation is kept in the form:

$$R\hat{x} = z \tag{30}$$

R is the upper triangular square-root of the information array and R and z are related to the normal equations as:

$$R^T R = H^T W H + P_{ap}^{-1},$$ (31)

$$z = (R^T)^{-1} H^T W y$$ (32)

and the covariance P of the solution (inverse of the information array) is given by:

$$P = R^{-1} (R^{-1})^T$$ (33)

We separate observations for gravity field determination into disjoint time spans called data arcs. Two-day-long data arcs are typical. The parameters estimated in the arc-by-arc gravity solutions consist of arc-dependent local variables: spacecraft state, solar radiation pressure coefficients, etc., and global variables common to all data arcs: gravity coefficients, tide parameters, etc. Merging the global parameter portion of a sequence of data arc square root information arrays produces a solution equivalent to solving for a single set of global parameters plus independent arc-specific local parameters (Kaula 1966).

When solving for a large number of parameters, convergence is very sensitive to a priori values and uncertainties. If the spacecraft initial state is poorly known and a filter tries to solve for both the trajectory and a high-resolution gravity field at the same time, the iteration may never converge. In order to avoid this problem, the local parameters are first estimated, and once a solution is obtained, the global parameters are estimated.

For each spacecraft, the local parameters consist of the spacecraft initial state, the solar radiation pressure scale factor, two constant SRP scaling terms orthogonal to the spacecraft-to-Sun vector, fifteen periodic acceleration terms for every two hours, four inter-satellite range-rate measurement correction terms for every two hours, and constant Earth-based Doppler bias and drift rate. Local parameters are used to constrain non-gravitational effects and measurement biases and are chosen based on experience. The global parameters consist of three inter-satellite range-rate time-tag biases, degree 2 and 3 Love numbers, degree 2 and order 1 amplitudes of periodic tidal signature, Moon's mass (GM), and a 150×150 gravity field (approximately 23,000 parameters). The time-tag biases represent the offset between the DSN time and a KBRR time-tag derived from the spacecraft clock.

Due to the accumulation of spacecraft angular momentum, maneuvers for Angular Momentum Desaturations (AMD) take place periodically. AMD maneuvers disrupt the quiet environment for gravity measurement and break the arc of data to be processed. Since we expect maneuvers, and to avoid numerical noise limitations on trajectory integration, we postulate 2-day arcs in our simulations. As described in Park et al. (2012), for each 2-day arc, we first estimate and re-estimate local parameters for each arc until convergence. Having converged on local parameters, we then compute SRIF arrays containing both local and global parameters for each arc, combine, and estimate, re-compute, re-combine, re-estimate, repeating until convergence.

6 Modeling Parameters

The input parameters to the simulations of the GRAIL mission are discussed below, grouped in the categories of data noise, data coverage, data arcs, orbital parameters, dynamic errors, and kinematic errors. To show the types of issues the simulation team was addressing, Table 2 lists a summary of parameters relevant to the simulation results and our model confidence in each one.

Table 2 Confidence level in GRAIL parameters relevant to science simulations

Parameter	Assumption	Note	Confidence level
Orbit initial conditions	Orbit conditions and spacecraft alignment are favorable	Inclination and node differences between spacecraft match requirements (0.02°)	High
Instrument noise spectrum	Spectrum includes thermal noise and USO jitter	GRACE analysis and performance modified for GRAIL	High
DSN data amount and noise	Tracking coverage is sufficient and noise characterization valid, includes USO	8 hours per day per spacecraft. DSN noise of 0.05 mm/s at 10-s integration time	High
KBR data continuity	No hardware resets	Tested with 5-min gaps once per day; show no impact	High
Time tag offset between payload and DSN time	Known to 100 ms, stable to 100 micro-seconds over 2 days	Convergence confirmed in science simulation	Medium to high
Temperature of spacecraft and payload elements	Linear dependence on beta angle	Tested conservatively, small error contribution	High
Propellant leakage	Constant and small	Preliminary information from spacecraft team	Medium
Outgassing	Small after cruise	Data from previous spacecraft	High
Lunar surface radiation	From lunar mission experience	Published models	Medium
Fuel slosh	Very small	Use of a propellant tank diaphragm	High
Solar radiation pressure	Constant reflectivity properties per arc and un-modeled errors <2 %	Currently investigating variability over an arc	Medium to high
Lunar librations	Modeled with Lunar Laser Ranging data	Known to a few milliseconds of arc	High
Lunar core signature	Monthly periodic	Phase not known	Medium

6.1 Data Noise

GRAIL simulations tools create DSN Doppler and inter-spacecraft Ka-band range rate data and apply noise to both data types. The simulations do not include DSN range data since this data type does not significantly improve GRAIL orbital accuracy. Since data noise levels are non-Gaussian, the applied noise and data weights assigned during follow-up parameter estimation are not always identical.

As discussed earlier, only the X-band Doppler link was included for simulation purposes, not the communications and navigation two-way S-band link between the DSN and the GRAIL spacecraft. The shorter wavelength X-band is less susceptible to ionosphere and interplanetary plasma noise. Expressed in units of velocity, our studies assume 0.05 mm/s DSN one-way X-band link residual noise and data weight at an integration time of 10 s, when simulating Doppler data and when filtering simulated data. This assumption is slightly more conservative than the typical noise level of 0.03 mm/s at 10 s experienced with the Mars Global Surveyor X-band performance.

The non-Gaussian residual noise associated with the payload's KBRR is added to the simulated inter-spacecraft data as a function of frequency. The long wavelength noise for 5-s samples is 0.4 µm/s for long wavelengths and then transitions to 1.0 µm/s at the short wavelength. However, in the filter, a constant data weight of 1 µm/s white noise is applied.

6.2 Data Coverage and Data Arcs

As a baseline, we simulated 8 hours of DSN daily tracking data for each spacecraft, non-overlapping, for a total of 16 hours. This DSN coverage provides information on absolute orbit for Ebb or Flow and improves long wavelength gravity field solutions, including the lunar core parameters, but contributes minimally to the global and regional science requirements. The coverage of the KBRR data is assumed to be continuous. Obtaining 16 hours of DSN coverage per day, every day, for one mission is considered very challenging due to the loading on the DSN but the requirements were accepted since the prime mission duration is relatively short, on the order of 3 months.

Since successive momentum dumps occur typically two days apart, GRAIL simulations assume a two-day data arc length, starting from an epoch of 4 March 2012 (actual epoch varied). Longer arcs are typically desirable but the momentum dumps are their natural boundaries.

6.3 Orbital Parameters

During the 82-day Science Phase, the Moon rotates three times underneath the GRAIL orbit. The collection of gravity data over one complete rotation, 27.3 days, is called one mapping cycle. Ebb and Flow are in a common near-polar, near-circular orbit with a mean altitude of approximately 55 km during the prime mission. However, as described in Roncoli and Fujii (2010) the periapsis altitude ranges from approximately 16 km to 51 km above a reference lunar sphere. The Ebb-Flow separation distance is designed to slowly vary. For approximately the first half of the mission, they drift apart and their separation distance increases from ~85 to ~225 km and then, with only one small orbit trim maneuver, they drift towards each other and the distance decreases to ~65 km near the end of the mission. The shorter separation distance is optimum for data exploring the local and regional spatial features while the segment around the maximum separation is optimum for the determination of the global studies such as the lunar core parameters, which are the Love number and the periodic signature of degree 2. The separation distances are designed to ensure that there is no degradation of the Ka-band signal due to multipath off the lunar surface and, according to Roncoli and Fujii (2010), a shorter spacecraft separation is required because of the lower spacecraft altitude.

The spacecraft separation contains a drift in order to reduce the resonance effects corresponding to the harmonic of the separation distance. Resonance effect degradation occurs at harmonic degrees of the form $N = 360/(D/30)$ where D is the spacecraft separation. This corresponds to degrees 54, 108, and 162 for a 200-km separation distance. For a 50-km separation, the resonance occurs much later, starting at degree 216.

The spacecraft inclination varies between approximately 88.4 and 89.85 degrees with a twice-per-month periodic signature. The average inclination, approximately 89.1 degrees, is offset from a perfectly polar orbit to improve the determination of low degree harmonic coefficients, but kept to a minimum to reduce the gap in data coverage at the poles.

The GRAIL science orbital phase is limited in part by a solar beta angle constraint of 49 degree imposed by the capability of the electrical power system; the spacecraft cannot generate sufficient power from the solar arrays for angles below this constraint. Figure 14 of Roncoli and Fujii (2010) illustrates the time history of the solar beta angle as well as the relationship between beta angle and the duration of solar eclipse during the science phase; eclipse durations are a maximum at the beginning and end and no solar eclipses when the solar beta angle is near 90 degrees near the middle. For the simulations the Sun-angle is represented after Park et al. (2012) with the inclination computed with respect to the lunar pole vector, as:

$$\beta = 90° - \cos^{-1}(\hat{e}_h \cdot \hat{r}_{sm})$$ (34)

To separate the non-gravitational signature from gravitational effects, a β angle of 90° is optimal. The spacecraft enters terminator crossing at $\beta \sim 76°$, and for β-angles less than this, modeling non-gravitational forces becomes more difficult, as the perturbations change rapidly due to partial shadowing of the orbit.

6.4 Dynamic Errors

The MIRAGE filter estimates both local parameters that are dependent for each data arc and global parameters that are common to all data arcs. The local dynamic parameters that are estimated include three dimensionless parameters for the solar pressure model of each spacecraft, a constant scale factor for the force along the sun-spacecraft direction with a nominal value of 1.0 and an a priori uncertainty of 0.10 or 10 % of the solar pressure force, and a scale factor for each off normal directions with a nominal value of 0.0 and an a priori uncertainty of 0.02 or 2 % of the overall solar pressure force. The solar pressure is modeled as a box-wing plate model with appropriate specular and diffuse coefficient values for each plate. Since both spacecraft are nearly identical, the solutions for the solar pressure scale factors are expected to be nearly equal. Several gravity solutions were also generated with strong a priori correlations between the two spacecraft solar pressure solutions to force them to be nearly equal. This constraint improved the core parameter uncertainties by about 30 % but had little effect on the global and regional gravity requirements. The baseline approach, however, is to treat the solar pressure solution of each spacecraft independently.

To account for un-modeled residual solar pressure errors, 15 periodic coefficients are estimated for each spacecraft for each arc with a priori amplitude equal to $\sim$2 % of the solar pressure force, or 3×10^{-12} km/s^2. The coefficients include constant, once per revolution and twice per revolution amplitudes for the radial, orbit normal, and along the velocity directions.

The lunar albedo model is not part of the baseline simulations results but albedo errors were independently investigated and found to be minimal. The albedo surface representation is given by the 10th degree spherical harmonic expansion of Floberhagen et al. (1999). The lunar surface thermal re-radiation is also investigated and is similar in size to albedo. Another non-gravitational force to be considered for GRAIL is the thermal radiation force as a result of heating on the spacecraft; this force is assumed to be small.

In the estimation of the gravity field, we assume a nominal gravity field and a truth gravity field. For the early simulation, the lunar gravity model derived from the Lunar Prospector mission to degree and order 150 and designated LP150Q in Konopliv et al. (2001) was used as both the nominal and truth models. Simulations since then have also used the smaller LP100J lunar gravity model to degree 100 as the nominal model to test convergence to the LP150Q truth model for different modeling assumptions.

The global dynamic parameters that are estimated include a gravity field to a given degree, the second-degree Love numbers, and the periodic amplitudes of C_{21} and S_{21} for core detection. In order to reduce computation time, the globally estimated gravity field was to degree 150 and extrapolated the results to degree 180. With current assumptions about data quality, it is expected that the gravity field will be recovered to higher than degree and order 300, which would make the Moon the body with the highest known gravity resolution in the universe.

6.5 Kinematic Errors

In addition to dynamic errors, which directly affect the spacecraft orbit because of a force, there are kinematic errors that affect the KBRR or DSN Doppler observables directly. Most of the kinematic corrections are estimated as local parameters that affect only that data arc. For every GRAIL orbit period of ∼2 hours, a KBRR bias, drift, and cosine and sine once per revolution amplitudes are estimated. For a 2-day arc this amounts to 48 parameters being estimated. No a priori constraint is applied to these parameters. For each arc, one DSN Doppler bias and one drift parameter are estimated to correct the USO frequency for the X-band one-way Doppler.

The very important kinematic error of timing offsets is addressed in a separate section below. Other kinematic errors are investigated by introducing noise or systematic trends in the KBRR residuals. The KBRR noise spectrum includes error contributions from the USO and spacecraft attitude jitter and the simulation account for them with a spectrum ranging from 0.4 µm/s at the long-wave-length to 1 µm/s at the short wavelength. Errors due to the offset of the phase center from the line connecting the center-of-mass from each spacecraft are minimized by actively pointing the spacecraft to align the phase centers. The changes in the temperature of the payload antenna and related hardware are modeled as a systematic trend in the observables. The current models we have investigated are sinusoidal once per orbit tones of 80 µm, a twice per orbit tone, and a triangular shaped twice per orbit signature. Amplitudes of all of these depend on the Sun angle. There is also a small error due to a shift reactive to the Ka-band system. These errors have a small effect on the core parameters and a negligible effect on the global and regional requirements. Errors due to sloshing of the fuel in the spacecraft fuel tank are expected to be negligible.

7 Modeling System Contributions

GRAIL utilized an integrated scientific measurement system comprised of flight, ground, mission, and data system elements in order to meet the end-to-end performance required for achieving the scientific objectives. Simulations leading to end-to-end error budgets were used to optimize the design, implementation, and testing of these elements. Because the inter-spacecraft range and range-rate observables and the range-rate between each spacecraft and ground stations can be affected by the performance of all elements of the mission, they were all treated as parts of an extended science instrument or a science system.

7.1 Flight System

The flight system is the spacecraft, which is comprised of the orbiters and science payload. The orbiter is comprised of the spacecraft bus, heritage from the Experimental Satellite System 11 (XSS-11) mission, and subsystems described in Hoffman (2009). At the heart of the flight system is the LGRS payload. Its design and performance are detailed in Klipstein et al. (2013, this issue). Key components of the GRAIL payload, such as the USO and RSB, are discussed here in the context of the quality of radio links, Doppler data, timing effects, and end-to-end error budgets.

As discussed above, to assess the ability of the flight system to meet the science requirements, all known sources of dynamic and kinematic error were assessed for their possible contributions to the uncertainty of simulated solutions for the gravity field and time varying tidal and core parameters. Fahnestock et al. (2012) described how early calculations of these effects revealed that the acceleration in reaction to spacecraft thermal re-radiation was among the dominating sources of solution uncertainty. Thus detailed modeling of the spacecraft's geometry, surface material, optical properties and thermal state was investigated along with approaches to the interface of such modeling with the science data processing.

First, a model of each spacecraft was constructed with plate surfaces defined for each spacecraft to cover the external area, not accounting for protuberances such as the thrusters and cameras. The two spacecraft are virtually identical but externally mirrored about the body-frame X–Z plane (see Fig. 4). Thermal modeling was performed by Lockheed Martin using the Thermal Desktop software, and for each spacecraft for the thermal loading conditions of varying beta angles. This included incident sunlight, lunar albedo and infra-red (IR) emission, and equipment power modes during the data acquisition period. Steady-state nodal temperature output, repeatable from orbit to orbit, was averaged over the plate surfaces to produce averaged temperature profiles for each plate surface over one orbit. These were used with the best available material optical properties, most measured after thermal vacuum bake-out, to compute body frame acceleration profiles over one orbit. These were then double interpolated over phase angle to obtain body frame acceleration at any epoch in the science mission.

As detailed in Fahnestock et al. (2012), the sensitivities of the thermal re-radiation acceleration history model with respect to assumptions and inputs were examined for impact on science. These included quantifying the difference spatial averaging of nodal temperature output versus no spatial averaging over the solar array surfaces, the sensitivity to extracting power from the solar arrays, the sensitivity to 3σ variations in the optical properties of every type of used material, and sensitivity to additive worst-case temperature biases on all surfaces.

A 5 K global bias bounded most changes resulting from realistic input variations. The difference in predicted thermal re-radiation acceleration history between that case and the nominal case was taken as indicative of the magnitude of un-modeled accelerations that would act on the spacecraft, which was used to create an a priori error model for the constant, and once and twice per orbit sinusoidal, periodic acceleration parameters in the simulations.

For the gravity field solution, the modeled thermal re-radiation acceleration history, a telemetry-derived thermal re-radiation acceleration history, or a hybrid of the two was included in the nominal dynamical model. For the first, the solar array bus open circuit voltage and short circuit current telemetry channels were utilized to determine the actual epoch of transit into and out of the Moon's shadow, and then the modeled history was shifted and stretched in time to match these transit times. For the telemetry-derived history, given n surfaces in the spacecraft model, each of $m < n$ instantaneous surface temperatures was set equal to the average of instantaneous readings of one or more temperature sensors, selected based on geometric proximity to that surface. Each of the $n - m$ remaining surface temperatures was set equal to the closest associated one of the m surface temperatures, plus a bias computed as the time average, over one orbit at $\beta = 90°$ epoch, of the difference of the pre-flight thermal modeling output for the two surfaces in question. The $\beta = 90°$ epoch had no eclipsing and was when the spacecraft were the most thermally and dynamically quiet, so we chose to tailor our mapping from sensor temperatures to surface temperatures to this time period. The hybrid history, illustrated in Fig. 5, was a combination of the modeled history for the Y and Z body axes directions, and the telemetry-derived history for body X axis direction. The apparent best methodology to use in data processing appears to be the hybrid thermal re-radiation acceleration history and a priori error model derived from it, as described earlier.

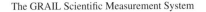
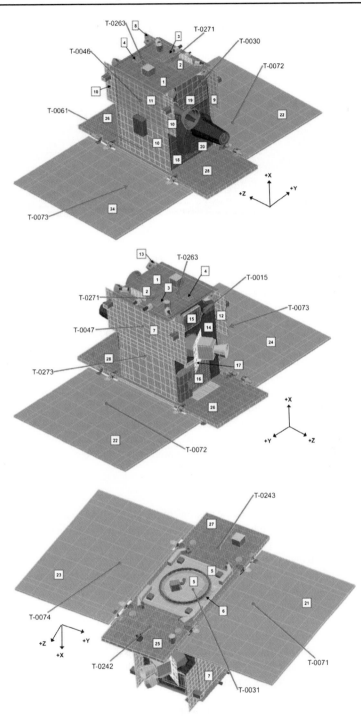

Fig. 4 Three views illustrating the placement of surfaces (*numbered in boxes*) on Flow's highest fidelity thermo-optical model, used for thermal re-radiation acceleration calculation. Ebb is similar, and temperature sensor locations are also noted

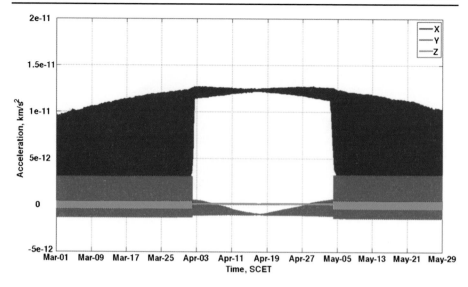

Fig. 5 Spacecraft body-frame components of hybrid thermal re-radiation acceleration history

The SRP acceleration was also included in the dynamical modeling in MIRAGE along with lunar albedo and IR emission pressure accelerations. For computing all three, a five-plate model of each spacecraft was obtained from the highest resolution model. This was done through combination of surfaces, first for common material and orientation, then for only common orientation, but always in a manner that kept the models roughly equivalent in four quantities: power coming into the spacecraft, acceleration due to SRP, power coming out of the spacecraft, and acceleration due to thermal re-radiation. All modeling of the non-gravitational accelerations within the spacecraft dynamics was sufficiently accurate to improve the orbit determination and data calibration as well as the gravity field solution.

7.2 Ground System

The ground system is comprised of the stations of the DSN, which was used exclusively by GRAIL, and the infrastructure facilities at JPL that support the DSN operations and transfer of data to users. The DSN's primary service to GRAIL and other missions is the telecommunications aspect, telemetry and commanding as well as navigation radio-metric data types. DSN services relevant to science are those providing precision measurements of the signal carrier frequency/phase for the purpose of Doppler observables. Two DSN subsystems Tracking and Radio Science are examined. As described earlier, GRAIL utilizes two bands for links to the DSN, a two-way S-band for telemetry, command, and navigation and a one-way X-band un-modulated carrier from the payload's RSB for science. As a result, DSN stations capable of S-band uplink and downlink as well as X-band downlink are required to support the mission, a criterion that narrows the available stations to a few 34-m diameter stations throughout the network but still enables the mission to get required coverage.

The Tracking Subsystem The Doppler data observable is generated in real-time at the DSN stations. The tracking receiver is a closed-loop system that finds the carrier frequency via a built-in algorithm and tracks it, aided by a prediction file for initial acquisition, producing the receiver's one-time real-time computation of the frequency and Automatic Gain Control

(AGC). Within its design threshold for dynamic conditions and signal-to-noise ratio, the output of the receiver is useful with a quantifiable Doppler noise that ranges between 0.02 and 0.1 mm/s. For GRAIL simulations, the X-band, USO driven link is associated with Doppler noise of 0.05 mm/s. If the threshold is exceeded, the receiver loses lock and data are not recoverable.

Radio Science Receiver   Designed specifically for Radio Science experiments described in an overview by Asmar (2010), the DSN's Radio Science Receiver (RSR) is at the heart of an open-loop reception/recoding subsystem that preserves the raw qualities of the electromagnetic wave propagating from the spacecraft source to the DSN stations. The digital receiver neither locks onto the carrier signal nor makes real-time decisions about its frequency or amplitude. Instead, it down-converts the signal in a predictions-driven heterodyne method and records the raw complex samples into files for users' post-pass processing. Asmar et al. (2005) describes the RSR usage and typical performance derived from other missions.

The use of the RSR proved to be critical for enhancing the quantity and quality of GRAIL X-band Doppler data in two ways: (1) practically double the amount of RSB data received by the DSN by an unofficial use of the concept of multiple-spacecraft per aperture, where the DSN station scheduled to track Ebb, for example, also views Flow and vice versa, and (2) contribute to understanding the various timing effects, as explained at length in Sect. 8. GRAIL funded the development of 3 portable RSRs for use throughout the DSN in support of GRAIL data acquisition and timing synchronization.

7.3 Mission System

The mission operations system is comprised of the JPL and DSN infrastructure as well as the Lockheed Martin operations. Since the mission design affects science data quality, the following factors had to be very carefully considered: spacecraft altitude and spatial variation of the altitude, spacecraft separation distance, orbit inclination, ground track separation, mission duration, number of maneuvers, time separation of maneuvers, AMD separation, and amount of ground station coverage. Mission design, navigation, deep space stations, and the ground data system critically contribute to the quality of science data. Specifics of each factor were described above and additional details on the mission system can be found in Roncoli and Fujii (2010), Hatch et al. (2010), Beerer and Havens (2012), and Zuber et al. (2013, this issue).

7.4 Data System

The GRAIL Science Data System (SDS) is comprised of all project hardware and software tools that contribute to the quality of the science data. Sine the SDS team is the first team to assess the quality of the data on a daily basis, it provides immediate feedback to the Mission System on the health of the spacecraft, payload, or ground system in case action is required to address an anomaly.

Zuber et al. (2013, this issue) provides a functional block diagram of the SDS that shows the data flow from all the sources to the final science users and the archives of the Planetary Data System. For science data processing, delivery, and archiving, the SDS is organized to provide daily support including weekends in order to handle the data volume as well as prevent any oversight of anomalies for any extended period of time.

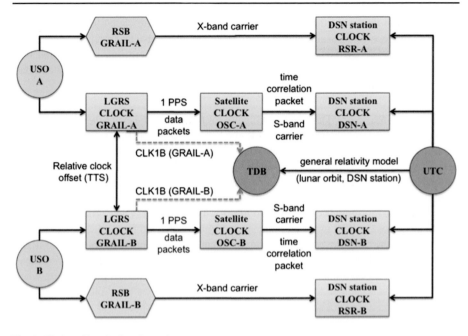

Fig. 6 Timing effects in the science data system

8 Data Timing and Synchronization

We have described how the process of determining the lunar gravity field starts with the inter-spacecraft Ka-band phase measurements used to compute the DOWR observables that are in turn converted to instantaneous range, range rate and range acceleration measurements. Very accurate timing of the measurements is crucial to achieve the high accuracy gravity results. The timing of the measurement has two components: absolute timing knowledge allows us to assign a measurement to a position around the Moon, and the relative timing knowledge between the payload clocks needed for the proper alignment of the LGRS phase measurements of both spacecraft at a common coordinate time. Following the formulation used in Kim (2000) and Thomas (1999), our analysis shows the aggregate errors in range to be below 1 μm and in range-rate to below 1 μm/s.

The GRAIL data are time-stamped on each spacecraft by the payload with time derived from the payload clock, namely the USO; Ebb and Flow each carries one USO that drifts independently from the other. The data are then passed to the spacecraft's central computer for packetizing prior to transmission to Earth as telemetry, and the computer puts a time-stamp on the packets derived from the spacecraft clock, which is independent from the USO. Finally, the data packets are received by the DSN and time-stamped at arrival, one-way light time after transmission, with the DSN time, which is derived from yet another independent clock. Counting two USOs, there are four independent clocks to synchronize in the post-processing in order to prevent errors in the gravity field solution, and this is carried out by the SDS team in the Level 1 processing stage as outlined in Fig. 6. The SDS team estimates the necessary time tag correction by combining information from available sources: the telemetry packets received at the DSN, the absolute frequency observed at the RSR, and the synchronization from the onboard inter-spacecraft TTS (note that additional observations were obtained at the DSN' RSR after launch of the inter-spacecraft TTS signals edge to deduce the absolute time on-board and correlate it to the known USO drift resulting in micro-second level of accuracy, orders of magnitude better than time correlation packet information).

For the purpose of the simulations, it is assumed that the a priori clock offset knowledge is 100 milliseconds, constant over one month, and that the reconstruction of the time tag offset is $\sim$20 microseconds. To achieve such accuracies, the RSB was added to the science payload to transmit a one-way X-band un-modulated sine-wave signal generated by the USO and recorded by the DSN's RSR. The RSR measurement of the frequency bias is $<10^{-5}$ Hz and standard deviation $= 10^{-3}$ Hz.

The DSN clocks are synchronized with the highly stable Coordinated Universal Time (UTC) standard. GRAIL's data processing, on the other hand, utilizes Barycentric Dynamical Time (TDB). The timing analysis derives a time correlation between the LGRS clocks and the TDB time scale, to be provided as a Level 1 ancillary data product called CLK1B. To produce CLK1B relating LGRS and TDB, we preprocess the timing data types and run them through a non-causal Kalman filter.

The LGRS clock on each spacecraft is driven by the USO for maximum stability, which is 3×10^{-13} over integration times of 1 to 100 seconds, expressed in Allan Deviation. This clock, however, does not report the absolute time but reports readout with respect to the clock startup epoch with errors from the drift of the USO. Relying on the nearly quadratic behavior of the onboard clock and an assessment of relativistic contributions, we believe that this system enables determining the relative time on Ebb vs. Flow with a bias $< 10^{-7}$ s and standard deviation $= 9 \times 10^{-11}$ s.

The Onboard Spacecraft Clock (OSC) is derived from a crystal oscillator with inferior stability to the USO. The onboard computer tags LGRS timing data packets with OSC time, including the LGRS 1 Pulse Per Second (PPS) packet. Ebb and Flow transmit time correlation packets to DSN stations where the arrival time is recorded in UTC, which provides a time correlation between the OSC and UTC. The DSN uses very stable hydrogen maser clocks and time-stamps the arrival of telemetry and tracking data in the UTC frame, which is tied to the International Atomic Time (TAI) frame. Based on DSN monitoring reports, the real-time timing performance of DSN time-tags is at the microsecond level and post-processing analysis improves the performance to the 10^{-9} second level.

By combining the LGRS/OSC and OSC/UTC time correlation products, a time correlation between LGRS time and UTC can be determined and the OSC clock drops out. Because OSC error is under one microsecond over intervals shorter than one second, the stability characteristics of the OSC do not limit LGRS and UTC correlation accuracy. Considering possible unknown timing delays in packet transmission, we expect a measurement bias of up to 100 milliseconds, and standard deviation of up to 30 milliseconds.

9 Relativistic Effects

Turyshev et al. (2013) has developed a realization of astronomical relativistic reference frames in the solar system and its application to the GRAIL mission. A model was developed for the necessary space-time coordinate transformations for light time computations addressed practical aspects of the implementation and all relevant relativistic coordinate transformations needed to describe the motion of the GRAIL spacecraft and to compute observable quantities. Relativistic effects contributing to the double one-way range observable, which is derived from one-way signal travel times between the two GRAIL spacecraft were accounted for and a general relativistic model for this fundamental observable of GRAIL, accurate to 1 μm and range-rate to 1 μm/s were also developed. The formulation justifies the basic assumptions behind the design of the GRAIL mission and may also be used in post-processing to further improve the results after the mission is complete.

It was recognized early during GRAIL's development phase that due to the expected high accuracy of ranging data, models of its observables must be formulated within the framework of Einstein's general theory of relativity in order to avoid significant model discrepancy. The ultimate observable model must correctly describe all the timing events occurring during the science operations of the mission for the links to Earth as well as the inter-spacecraft links. The model must take into account the different times at which the events have to be computed, involving the time of transmission of the Ka-band signal at one of the spacecraft, say Ebb, at the reception of this signal by its twin, Flow. In addition, the model must include a description of the process of transmitting S-band and X-band signals from both spacecraft and reception of this signal at a DSN tracking station.

Relevant points regarding relativistic corrections at the level of accuracy required by GRAIL include: (1) for a spacecraft around the Moon, we can model proper time treating the Moon as a point mass; (2) JPL's long-standing ODP models designed for proper time of a station on Earth are already sufficiently accurate, with no changes required; (3) up to a constant bias, computing one-way range from DOWR requires a pair of corrections from one-way light time to instantaneous distance. It suffices to iteratively solve for light time in terms of instantaneous distance, by re-computing transmission position bearing in mind the elapsed light time, in the presence of the Shapiro delay.

Turyshev et al. (2013) also notes that measuring the signal frequency involves computing three numbers: the derivative of proper time at the receiver with respect to coordinate time of reception, the derivative of proper time at the transmitter with respect to coordinate time of transmission, and the derivative of coordinate time of reception with respect to coordinate time of transmission. The first number must be modified to account for the fact that the clock at the DSN receiver attempts to synchronize with UTC time, rather than simply acting as a TDB receiver placed on the surface of the earth. The effect of the Earth's and the Moon's gravity on the third term will be below our level of error; if we did choose to include them it would certainly suffice to use a point mass.

10 Results of Simulations

10.1 A Priori Assumptions and Kaula Constraints

The a priori uncertainties for the models used in the simulations are summarized in Table 3. Furthermore, we assume that the KBRR data have $\sigma_s = 1$ μm/s uncertainty at 5-second count time, and the DSN Doppler tracking uncertainty at 10-second count time $\sigma_d = 0.05$ mm/s, obtained from previous payload and flight experience. The KBRR data give an average accuracy of 2×10^{-10} km/s², which is equivalent to 0.002 mGal in the gravity measurement.

Gravity field estimates often require a constraint by the Kaula rule (Kaula 1966). In this study, we assume the Kaula constraint to be $2.5 \times 10^{-4}/n^2$. Note that the Kaula constraint approximates the root-mean-square (RMS) of the spherical harmonics coefficients, i.e.,

$$\text{RMS}_n = \sqrt{\frac{\overline{\sigma}_n^2}{2n + 1}}, \tag{35}$$

for the degree variance $\overline{\sigma}_n^2 = \sum_{m=0}^{n}(\overline{C}_{nm}^2 + \overline{S}_{nm}^2)$.

Table 3 GRAIL simulations a priori uncertainty model

Parameters	A priori uncertainty
Position	1000 km in each direction
Velocity	1 m/s in each direction
Overall SRP scaling factor	10 %
Orthogonal SRP scaling factors	2 % in each direction
Doppler correction terms	open
LGRS correction terms	open
Periodic acceleration	Variable
Lunar gravity field	LP150Q solution
k_2, α_{21}, and β_{21}	open

Table 4 Baseline core-signature estimates

	$\Delta_{k2} \times 10^4$	$\sigma_{k2} \times 10^4$	$\Delta_{\alpha 21} \times 10^{10}$	$\sigma_{\alpha 21} \times 10^{10}$	$\Delta_{\beta 21} \times 10^{11}$	$\sigma_{\beta 21} \times 10^{11}$
Requirement		2.00		1.00		2.50
Baseline	0.93	0.94	0.62	0.67	−2.01	1.77

10.2 Dynamic Conditions and Core Signature Model Results

Table 4 shows the estimates of the core signature terms for the baseline case, where Δ_i represents the difference between the truth and the estimated parameter i and σ_i represents the estimated 1-σ uncertainty of parameter i. The effect of different data arc lengths, which are bounded by spacecraft maneuvers with a baseline of 2 days, is that the longer the arc the better the results for the core signature, as expected (Park et al. 2012). The baseline assumes Ebb and Flow are tracked by the DSN for at least 8 hours per spacecraft per day. Longer tracking is better and the diversified coverage over different DSN complexes provides additional orbit information due to geometric parallax.

The remaining possible error contributions from un-modeled non-gravitational accelerations are applied as an a priori periodic acceleration model that impacts the formal uncertainties of the estimated low-degree gravity field and time-varying core signature. The minimum periodic acceleration is chosen to be 3×10^{-13} km/s^2, as shown in Fig. 7, and scaled up accordingly for the period of lower spacecraft separation; with this minimum a priori periodic acceleration, all science requirements are met. We showed in Park et al. (2012) that varying the a priori periodic acceleration to the level of 1×10^{-12} km/s^2 still allows satisfying all science requirements but with smaller error margin for the core parameters.

10.3 Kinematic Error Results

Table 5 shows the effect of kinematics errors on the estimated core parameters. The error in the estimated parameters represents the effect due to kinematic errors and the formal uncertainties are the same as in the baseline case. The temperature control case shows the effect of the error in the LGRS measurement frequency and signal path-length due to thermal variation. The time-tag error shows the contribution to the total error of estimating the KBRR time tag with an initial time-tag offset of 100 milliseconds. Lastly, the attitude pointing error shows the effect of a 3-σ single-spacecraft attitude pointing error, which translates to about 0.06 µm/s on LGRS data.

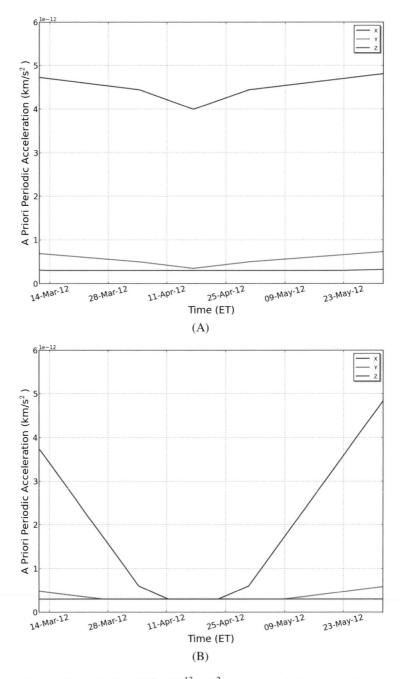

Fig. 7 A priori periodic acceleration with 3×10^{-13} km/s^2 minimum acceleration constant term in panel (**A**) and once-per-orbit and twice-per-orbit terms in panel (**B**) (Park et al. 2012)

Table 5 Effect of kinematics errors on estimated core parameters

Cases	$\Delta_{k2} \times 10^4$	$\Delta_{\alpha21} \times 10^{10}$
Temperature control	0.05	0.01
Time-tag error	0.02	0.02
Attitude pointing error	0.20	0.10

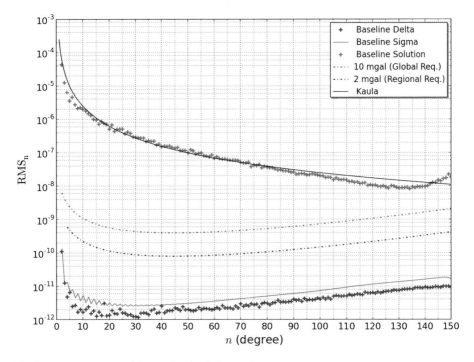

Fig. 8 Root mean square of the baseline simulation results

10.4 Summary

The most significant result of the simulations is the RMS of the estimated gravity field corresponding 1-σ formal uncertainties and the difference between the truth and estimated gravity fields, shown in Fig. 8. Also shown in the same figure are the global and regional science requirement lines generated based on the surface acceleration accuracy required for global and regional science requirements, which are satisfied with the baseline assumptions. The difference between the truth and estimated gravity fields is smooth and is bounded by the formal uncertainty indicating a correct filter setup and a stable filter solution. The colored measurement noise is well bounded by the white noise assumption that was used in the estimation process since the recovered values are well represented by the formal uncertainties. The linear extrapolation of the estimated uncertainties indicates that a nominal gravity field of degree 300 or better can be determined according to the Kaula rule. The largest source of non-gravitational error comes from spacecraft thermal radiation, which is characterized with variable a priori error constraint model derived from orbit geometry and expected force magnitude.

With all error models included, detailed and numerous simulations show that estimating the lunar gravity field is robust against dynamic and kinematic errors and meets the high accuracy lunar gravity requirements by at least an order of magnitude. A nominal lunar gravity field of degree 300 or better can be achieved according to the scaled Kaula rule for the Moon. The core signature is more sensitive to modeling errors and depends on how accurately the spacecraft dynamics can be modeled; the requirement can be achieved with a small margin.

Acknowledgements   The GRAIL mission is supported by the NASA Discovery Program under contracts to the Massachusetts Institute of Technology and the Jet Propulsion Laboratory. The work described in this paper was mostly carried out at Jet Propulsion Laboratory, California Institute of Technology, under contract with the National Aeronautics and Space Administration. The authors thank colleagues who have contributed to this work or reviewed it, especially at JPL: Duncan McPherson, Ralph Roncoli, William Folkner, Kevin Barltrop, Charles Dunn, William Klipstein, Randy Dodge, William Bertch, Daniel Klein, Dong Shin, Stefan Esterhausin, Slava Turyshev, Tom Hoffman, Charles Bell, Hoppy Price, Neil Dahya, Joseph Beerer, Glen Havens, Robert Gounley, Ruth Fragoso, Susan Kurtik, Behzad Raofi, and Dolan Highsmith. From Lockheed Martin Space Systems Company (Denver): Stu Spath, Tim Linn, Ryan Olds, Dave Eckart, and Brad Haack, Kevin Johnson, Carey Parish, Chris May, Rob Chambers, Kristian Waldorff, Josh Wood, Piet Kallemeyn, Angus McMechan, Cavan Cuddy, and Steve Odiorne. From the NASA Goddard Space Flight Center: Frank Lemoine and David Rowlands, and from the University of Texas: Byron Tapley and Srinivas Bettadpur.

References

S. Aoki, H. Kinoshita, Note on the relation between the equinox and Guinot's non-rotating origin. Celest. Mech. **29**, 335–360 (1983)

S. Aoki, B. Guinot, G.K. Kaplan, H. Kinoshita, D. McCarthy, P.K. Seidelmann, The new definition of universal time. Astron. Astrophys. **105**, 359–361 (1982)

D.F. Argus, R.G. Gordon, No-net-rotation model of current plate velocities incorporating plate motion model NUVEL-1. Geophys. Res. Lett. **18**, 2039–2042 (1991)

S.W. Asmar, Radio as a science tool. Proc. IEEE **98**, 10 (2010)

S.W. Asmar, J.W. Armstrong, L. Iess, P. Tortora, Spacecraft Doppler tracking: noise budget and achievable accuracy in precision radio science observations. Radio Sci. **40** (2005). doi:10.1029/2004RS003101

J.G. Beerer, G.G. Havens, Operation the dual-orbiter GRAIL mission to measure the Moon's gravity, in *SpaceOps 2012 Conference*, Stockholm, Sweden, June 2012

C. Boucher, Z. Altamimi, L. Duhem, Results and analysis of the ITRF93. IERS Technical Note, 18, Observatoire de Paris, 1994

C. Dunn, W. Bertiger, Y. Bar-Sever, S. Desai, B. Haines, D. Kuang, G. Franklin, I. Harris, G. Kruizinga, T. Meehan, S. Nandi, D. Nguyen, T. Rogstad, J.B. Thomas, J. Tien, L. Romans, M. Watkins, S.C. Wu, S. Bettadpur, J. Kim, Instrument of GRACE: GPS augments gravity measurements. GPS World **14**, 16–28 (2003)

E.G. Fahnestock, Comprehensive gravity and dynamics model determination of binary asteroid systems, in *American Astronomical Society, DPS Meeting 41, #50.11* (2009)

E.G. Fahnestock, R.S. Park, D.-N. Yuan, A.S. Konopliv, Spacecraft thermal and optical modeling impacts on estimation of the GRAIL lunar gravity field, in *AIAA/AAS Astrodynamics Specialist Conference*, Minneapolis, MN, AIAA, August 13–16, 2012, pp. 2012–4428

R. Floberhagen, P. Visser, F. Weischede, Lunar albedo forces modeling and its effect on low lunar orbit and gravity field determination. Adv. Space Res. **23**, 378–733 (1999)

W.M. Folkner, DSN station locations and uncertainties. JPL TDA Progress Report, 42-128, 1-34, Jet Propulsion Laboratory, California Institute of Technology, Pasadena, CA, 1996

W.M. Folkner, J.A. Steppe, S.H. Oliveau, Earth orientation parameter file description and usage. Interoffice Memorandum 335.1-11-93 (internal document), Jet Propulsion Laboratory, California Institute of Technology, Pasadena, CA, 1993

J. Guinn, P. Wolff, TOPEX/Poseidon operational orbit determination results using global positioning satellites, in *AAS/AIAA Astrodynamics Specialists Conference* (1993). AAS-93-573

S. Hatch, R. Roncoli, T. Sweetser, GRAIL trajectory design: lunar orbit insertion through science, in *AIAA Guidance, Navigation, and Control Conference*, Toronto, Ontario, Canada, August 2010. AIAA 2010-8385

W.A. Heiskanen, H. Moritz, Physical geodesy. Bull. Géod. **86**(1), 491–492 (1967)

T.L. Hoffman, GRAIL: gravity mapping the Moon, in *Aerospace Conference, 2009 IEEE*, Big Sky, MT, 7–14 March 2009. ISBN 978-1-4244-2622-5

W.M. Kaula, *Theory of Satellite Geodesy* (Blaisdell, Waltham, 1966). 124 pp.

J. Kim, Simulation study of a low-low satellite-to-satellite tracking mission. Ph.D. dissertation, Univ. of Texas at Austin, May 2000

J. Kim, B. Tapley, Error analysis of a low-low satellite-to-satellite tracking mission. J. Guid. Control Dyn. **25**(6), 1100–1106 (2002)

W.M. Klipstein, B.W. Arnold, D.G. Enzer, A.A. Ruiz, J.Y. Yien, R.T. Wang, C.E. Dunn, The lunar gravity ranging system for the gravity recovery and interior laboratory (GRAIL) mission. Space Sci. Rev. (2013, this issue)

A.S. Konopliv, S.W. Asmar, E. Carranza, D.N. Yuan, W.L. Sjogren, Recent gravity models as a result of the lunar prospector mission. Icarus **150**, 1–18 (2001)

F.T. Krogh, Changing stepsize in the integration of differential equations using modified divided differences. JPL Tech. Mem. No. 312, Section 914 (internal document), Jet Propulsion Laboratory, California Institute of Technology, Pasadena, CA, 1973

C.L. Lawson, R.J. Hanson, *Solving Least Squares Problems*. SIAM Classics in Applied Mathematics, vol. 15 (Society for Industrial and Applied Mathematics, Philadelphia, 1995)

R. Leavitt, A. Salama, Design and implementation of software algorithms for TOPEX/POSEIDON ephemeris representation, in *AIAA/AAS Astrodynamics Specialists Conference* (1993). AAS-93-724

J.H. Lieske, T. Lederle, W. Fricke, B. Morando, Expressions for the precession quantities based upon the IAU (1976) system of astronomical constants. Astron. Astrophys. **58**, 1–16 (1977)

A.J. Mannucci, B.D. Wilson, D.-N. Yuan, C.H. Ho, U.J. Lindqwister, T.F. Runge, A global mapping technique for GPS-derived ionospheric total electron content measurements. Radio Sci. **33**(3), 565–582 (1998)

D.D. McCarthy, G. Petit (eds.), *IERS Conventions*, IERS Technical Note, vol. 32 (2003)

T.D. Moyer, *Formulation for Observed and Computed Values of Deep Space Network Data Types for Navigation* (Wiley, Hoboken, 2003). 576 pp.

X.X. Newhall, J.G. Williams, Estimation of the lunar physical librations. Celest. Mech. Dyn. Astron. **66**, 21–30 (1997)

R.S. Park, S.W. Asmar, E.G. Fahnestock, A.S. Konopliv, W. Lu, M.M. Watkins, Gravity recovery and interior laboratory simulations of static and temporal gravity field. J. Spacecr. Rockets **49**, 390–400 (2012)

S. Pines, Uniform representation of the gravitational potential and its derivatives. AIAA J. **11**, 1508–1511 (1973)

R. Roncoli, K. Fujii, Mission design overview for the gravity recovery and interior laboratory (GRAIL) mission, in *AIAA Guidance, Navigation, and Control Conference*, Toronto, Ontario, Canada, August 2010. AIAA 2010-8383

P.K. Seidelmann, 1980 IAU theory of nutation: the final report of the IAU working group on nutation. Celest. Mech. **27**, 79–106 (1982)

B.D. Tapley, B. Schutz, G. Born, *Statistical Orbit Determination* (Elsevier, Boston, 2004a). 547 pp.

B.D. Tapley, S. Bettadpur, M. Watkins, C. Reigber, The gravity recovery and climate experiment: mission overview and early results. Geophys. Res. Lett. **31** (2004b). doi:10.1029/2004GL019920

B.J. Thomas, An analysis of gravity-field estimation based on intersatellite dual-1-way biased ranging. JPL Publication 98–15, May 1999

S.G. Turyshev, V.T. Toth, M.V. Sazhin, General relativistic observables of the GRAIL mission. Phys. Rev. D **87**, 024020 (2013)

J.M. Wahr, The forced nutations of an elliptical, rotating, elastic, and oceanless Earth. Geophys. J. R. Astron. Soc. **64**, 705–727 (1981)

R.C. Weber, P.-Y. Lin, E.J. Garnero, Q. Williams, P. Lognonne, Seismic detection of the lunar core. Science **331**, 309–313 (2011). doi:10.1126/science.1199375

J.G. Williams, A scheme for lunar inner core detection. Geophys. Res. Lett. **34**, L03202 (2007). doi:10.1029/2006GL028185

J.G. Williams, D.H. Boggs, W.M. Folkner, DE421 lunar orbit, physical librations, and surface coordinates. JPL IOM 335-JW, DB, WF-20080314-001, March 14, 2008

D.-N. Yuan, W. Sjogren, A. Konopliv, A. Kucinskas, Gravity field of mars: a 75th degree and order model. J. Geophys. Res. **106**(E10), 23377–23401 (2001)

M.T. Zuber, D.E. Smith, D.H. Lehman, T.L. Hoffman, S.W. Asmar, M.M. Watkins, Gravity recovery and interior laboratory (GRAIL): mapping the lunar interior from crust to core. Space Sci. Rev. (2013, this issue). doi:10.1007/s11214-012-9952-7

DOI 10.1007/978-1-4614-9584-0_4
Reprinted from *Space Science Reviews* Journal, DOI 10.1007/s11214-013-9973-x

The Lunar Gravity Ranging System for the Gravity Recovery and Interior Laboratory (GRAIL) Mission

**William M. Klipstein · Bradford W. Arnold ·
Daphna G. Enzer · Alberto A. Ruiz · Jeffrey Y. Tien ·
Rabi T. Wang · Charles E. Dunn**

Received: 3 October 2012 / Accepted: 1 March 2013 / Published online: 20 April 2013
© US Government 2013

Abstract The Lunar Gravity Ranging System (LGRS) flying on NASA's Gravity Recovery and Interior Laboratory (GRAIL) mission measures fluctuations in the separation between the two GRAIL orbiters with sensitivity below 0.6 microns/Hz$^{1/2}$. GRAIL adapts the mission design and instrumentation from the Gravity Recovery and Climate Experiment (GRACE) to a make a precise gravitational map of Earth's Moon. Phase measurements of Ka-band carrier signals transmitted between spacecraft with line-of-sight separations between 50 km to 225 km provide the primary observable. Measurements of time offsets between the orbiters, frequency calibrations, and precise orbit determination provided by the Global Positioning System on GRACE are replaced by an S-band time-transfer cross link and Deep Space Network Doppler tracking of an X-band radioscience beacon and the spacecraft telecommunications link. Lack of an atmosphere at the Moon allows use of a single-frequency link and elimination of the accelerometer compared to the GRACE instrumentation. This paper describes the implementation, testing and performance of the instrument complement flown on the two GRAIL orbiters.

Keywords GRAIL · Gravity · Moon · GRACE · Ranging

Acronyms

ADEV	Allan Deviation
ASIC	Application-Specific Integrated Circuit
DOWO	Dual One-Way (time) Offset
DOWR	Dual One-Way Range
DSN	Deep Space Network
EGSE	Electronic Ground Support Equipment
FS	Flight System
GPA	Gravity recovery Processor Assembly
GRA	GRAIL orbiter A ("Ebb")
GRB	GRAIL orbiter B ("Flow")

W.M. Klipstein (✉) · B.W. Arnold · D.G. Enzer · A.A. Ruiz · J.Y. Tien · R.T. Wang · C.E. Dunn
Jet Propulsion Laboratory, California Institute of Technology, Pasadena, CA 91109, USA
e-mail: Klipstein@jpl.nasa.gov

GPS   Global Positioning System
GRACE   Gravity Recovery and Climate Experiment
GRAIL   Gravity Recovery and Interior Laboratory
IF   Intermediate Frequency
KBR   Ka-band Ranging Assembly
LGRS   Lunar Gravity Ranging Assembly
MWA   MicroWave Assembly
PL   Payload
PRN   Pseudo-Random Noise
RPSD   Root Power Spectral Density
RSB   Radioscience Beacon
RSR   Radioscience Receiver
TTFE   Time Transfer Front End
TTS   Time Transfer System
USO   Ultra-Stable Oscillator

1 Introduction

The science goal of the GRAIL mission (Zuber et al. 2013) is to determine the gravitational field of the Moon to an accuracy sufficient to allow scientific questions about cratering processes, internal structure and thermal evolution of the Moon to be addressed. The approach taken by the mission to do this was to exploit a measurement technique in use by the Gravity Recovery and Climate Experiment (GRACE) mission (Tapley et al. 2004; Dunn et al. 2002) that is currently operating in Earth orbit. Like GRACE, the GRAIL observation consists of the time series of distance changes between two satellites in following polar orbits.

In GRACE and GRAIL the two satellites essentially form a single-axis gradiometer that measures the differential effect of gravity on test bodies separated by 50–250 kilometers. As shown in Fig. 1 the LGRS instrument involves three radiofrequency links from each orbiter. The primary measurement is achieved using a Ka-band carrier-only signal exchanged between the two orbiters. As on GRACE the phase of the transmitted and received signals from each orbiter are compared and combined on the ground to form the "Dual One Way Range" (DOWR) observable, described further below. An S-band Time Transfer System (TTS) cross link measures the time offset between the instruments with accuracy below 100 nanoseconds and provides a measurement of frequency changes in the on-board Ultra-Stable Oscillators (USOs) that form the radiometric instrument reference. An X-band Radioscience Beacon (RSB) transmits a carrier-only signal to be tracked by the DSN to calibrate the USO frequency on each orbiter; the RSB also provides one-way Doppler tracking to augment tracking of the telecom signal to Earth.

Gravity recovery relies both on the Ka-band DOWR as well as Doppler tracking of the S-band telecom system and RSB tone tracking by the Deep Space Network (DSN). This ground tracking of the spacecraft provides orbital constraints on the precise DOWR signal and contributes critically to the long-wavelength gravity recovery performance. The two-way telecom signal at S-band eliminates USO noise present in the RSB one-way signal, which exhibits lower ionospheric disturbances characteristic at X-band.

2 Instrument Overview

Figure 2 shows a block diagram of the hardware components of the payload complement on each orbiter. A USO provides reference signals to the Microwave Assembly (MWA), the

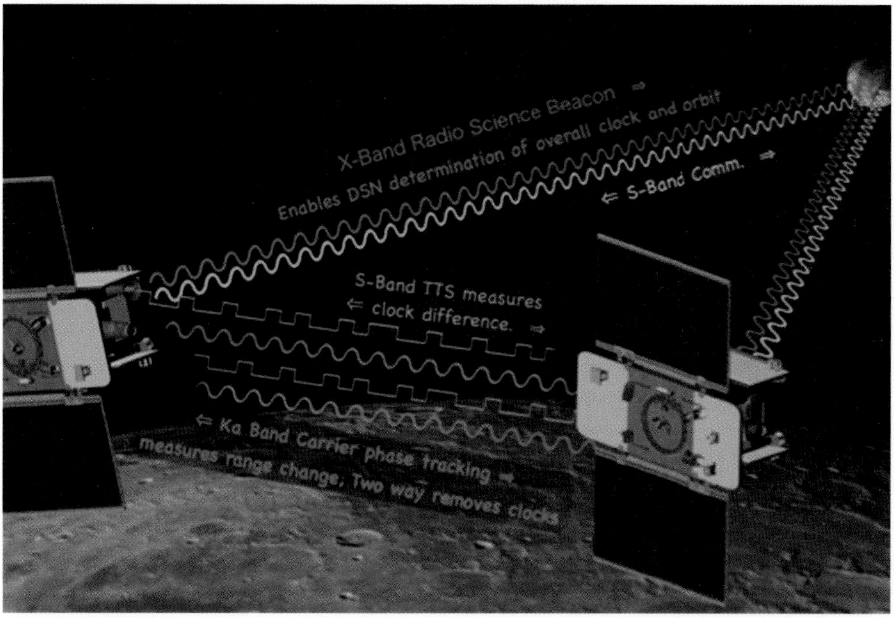

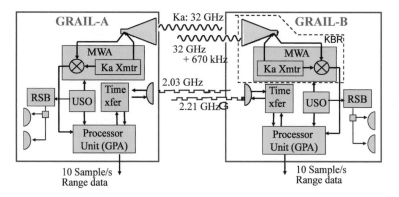

Fig. 1 GRAIL is a two-spacecraft mission that senses the Moon's gravity field by measuring changes in the separation between the orbiters with sub-micron precision

Fig. 2 Block diagram of the instrument complement on each orbiter. GRAIL-A and GRAIL-B were renamed "Ebb" and "Flow" respectively after launch

Gravity Processor Assembly (GPA), and the RSB following the frequency scheme shown in Table 1. This frequency scheme provides ultra-stable phase stability to all parts of the system from the outset. The MWA multiplies the USO signal up to Ka-band and transmits this signal through a microwave horn to the other orbiter. A portion of the Ka-band signal is also used as a local oscillator to mix with the signal received from the other orbiter to produce a baseband signal at approximately 670 kHz to be measured by the GPA. The MWA is connected to the microwave horn via waveguide in a Ka-band Ranging (KBR) assembly mounted to the exterior of the spacecraft. A radome attached to the horn keeps sunlight out of the horn to reduce thermally induced phase errors. The Ka-band transmit and

Table 1 The USO drives all radiometric elements coherently

	Ebb	Flow
USO Frequencies		
USO base frequency (MHz)	4.832000	4.832099
×8 = GPA, RSB inputs (MHz)	38.656000	38.656792
×12 = MWA input (MHz)	57.984000	57.985188
TTS Frequencies		
TTS Tx synth multiplier	$105/2 + 39333/2^{19}$	$114/2 + 48322/2^{19}$
TTS Tx (MHz)	2032.340041	2207.000021
TTS Rx freq (MHz)	2207.000021	2032.340041
TTS sample rate (MHz)	19.328	19.328396
Sample rate harmonic	114	105
Virtual LO (MHz)	2203.392	2029.48158
Carrier offset (MHz)	−3.608020707	−2.858460527
RSB Frequencies		
RSB synth multiplier ratios	$218 + 333540/2^{19}$	$218 + 333904/2^{19}$
RSB transmit freq (MHz)	8451.600061	8451.800059
MWA frequencies		
MWA multiplier	564	564
Ka Tx freq (MHz)	32702.976000	32703.646032
Ka-band IF (MHz)	0.670032	−0.670032

receive signals are separated by linear polarization in an orthomode transducer. The GPA measures fluctuations in the phase of the 670 kHz as the primary science observable with a precision below 10^{-4} cycles/Hz$^{1/2}$, yielding a resolution below 1 micron/Hz$^{1/2}$ based on the 0.0092-meter wavelength of the Ka-band carrier. The samplers on the GPA are clocked by a signal from the USO. The GPA also generates an S-band carrier coherent with the USO for the TTS transmit signals. Transmit and receive signals between the two orbiters are separated by frequency using a diplexer in the Time-Transfer Front End (TTFE) which connects to the Time Transfer Antenna. A third signal from the USO acts as a reference for the RSB, which synthesizes an X-band signal for transmission to Earth. The RSB signal passes through a switch connected to two antennas to provide visibility to Earth; the RSB antennas are switched by the spacecraft at the same time as the telecom antennas.

Figure 3 shows the hardware complement on each orbiter. Johns Hopkins Applied Physics Laboratory produced the USOs; Space Systems produced the MWAs, which are incorporated into the KBR and not visible in the figure; Custom Microwave, Inc. produced the waveguides and horn; JPL produced the GPA, the TTS, and the mechanical and thermal structure for the KBR. Figure 4 shows the mounting location of most of these elements on each of the spacecraft, Ebb and Flow.

The hardware on each spacecraft is nearly identical. The USOs differ by approximately 20 parts per million to produce the 670 kHz frequency offset in the downconverted Ka-band signals. The S-band TTS signals differ in frequency to allow separation of transmit and receive signals. The KBR mechanical and thermal designs differ due to the different spacecraft orientations and the approximately 2 degree pointing needed to point along the orbit chord between the nadir-pointed spacecraft. The two RSBs have a slightly different programmed frequency settable with an external connector.

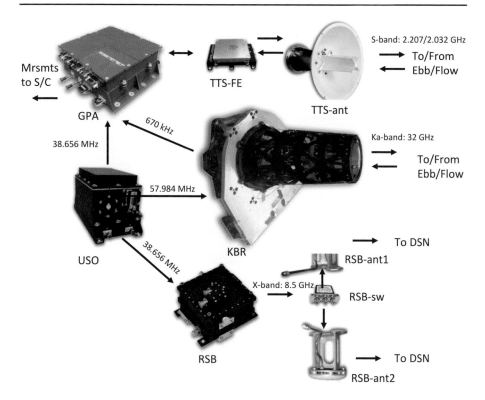

Fig. 3 LGRS flight hardware

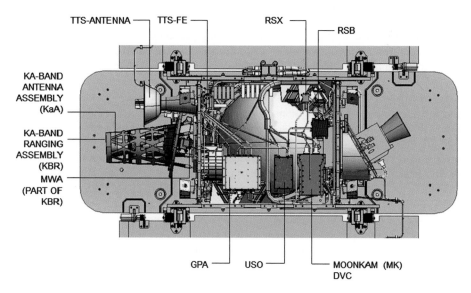

Fig. 4 GRAIL payload components shown on the spacecraft, viewed from opposite the solar panel normal. In addition to the LGRS components the payload included an Education and Public Outreach camera system ("MoonKAM") under the direction of Sally Ride Sciences

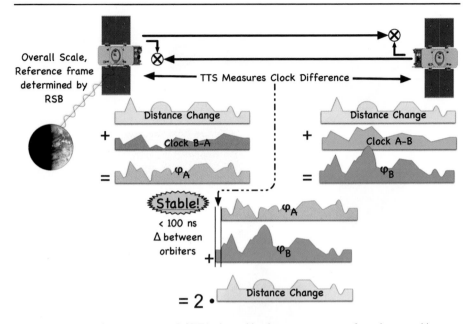

Fig. 5 In the Dual One-Way Range (DOWR) observable phase measurements from the two orbiters are summed on the ground to provide sensitivity to distance changes. A time transfer system allows data from the two orbiters to be lined up in time

The signal processing algorithms for the Ka-band signal are the same as those used on GRACE. Each spacecraft transmits a sinusoidal signal at Ka, with the two frequencies offset from each other by 670 kHz. At each satellite, the received Ka-band signal is down-converted to this baseband frequency using a local oscillator (LO) signal coupled from the corresponding Ka-band signal to be transmitted by that satellite. The antennae on each spacecraft are rotated 45 degrees around their boresight, so that in the flight configuration they form mirror images of each other. This allows orthogonal polarizations to separate the transmitted and received channels in the antenna.

The offset frequency, adopted from GRACE, was chosen to balance the need to reduce crosstalk between the send and receive Ka-band channels against the residual low frequency sensitivity to USO noise resulting from a non-zero offset frequency. In addition, the baseband phase measurement has been optimized for a non-zero baseband frequency. The 670 kHz down-converted Ka-band signals are sampled and passed to the digital signal processing (DSP) part of the LGRS. A dedicated channel in the receiver tracks the phase of the IF signal with a digital phase-locked loop, and extracts the phase using signal processing techniques inside the TurboRogue ASIC used in the BlackJack family of GPS receivers. The receiver outputs phase in cycles at 10 points per second, and these values are transmitted to ground for combining with the data from the other satellite and the formation of Ka-band DOWR, which is the primary instrumentation observable.

Figure 5 illustrates the DOWR observable. At each orbiter the mixed down phase includes information about the phase of the USO on each orbiter and fluctuations in the separation, which causes changes in the light time between the orbiters. Changes in the separation are correlated on the two orbiters; a longer light time for one certainly means a longer light time for the other. USO phase noise, in contrast, causes opposite changes on the two sides. The local oscillator mixing scheme results in the phases on the two orbiters counter-rotating, so an "advancing" phase causes an immediate positive shift at the transmitting spacecraft, and a negative shift a light-time later for the receiving spacecraft. The sum of the signals from the two orbiters thus gives twice the distance change, suppressing USO phase noise. See Thomas (1999) for more details.

In order to combine the data from each orbiter in post processing the time offset between samples on the two orbiters must be known. To achieve this, each satellite also transmits a pseudorandom noise (PRN) modulated spread-spectrum signal at S-band on the TTS (see Table 1 and Table 4). The PRN code transmitted across the link is modulated timing information in the form of a pseudo-range measurement, essentially a light time delay plus time offset. At each satellite, the received PRN modulated spread-spectrum signal from the other satellite is processed using the existing technology in the LGRS for GPS code generation inherited from the GRACE GPS receiver. A Frequency Division Multiple Access scheme is implemented to avoid self-jamming, where Ebb transmits at 2032 MHz, and Flow transmits at 2207 MHz. These frequencies are chosen to avoid interference between the TTS and the S-band spacecraft radio signals, which transmit at approximately 2280 MHz. TTS timing data at 1 Hz gets transmitted to the spacecraft as part of the instrument science data.

2.1 Comparison to GRACE

The most significant change in the GRAIL instruments is the removal of capability to calibrate error sources that primarily originate from the Earth's atmosphere and so are largely not present at the Moon. Specifically, GRACE carried a precision accelerometer to measure non-gravitational forces, while GRAIL relies on modeling. GRACE also included a coherent K-Band link to calibrate ionospheric effects, which are negligible at the Moon. GRACE relied on the Global Positioning System (GPS) for precision orbit determination and time coordination between the instruments on each orbiter. For GRAIL absolute orbit determination comes from DSN two-way Doppler tracking of the spacecraft telecom signal and the one-way X-band RSB signal. Time coordination was achieved using the S-band instrument crosslink.

2.2 Instrument Operations

The LGRS continuously measures the Ka-band and S-band cross links. Upon application of power the GPA on each spacecraft begins transmitting a pseudorandom noise code derived from the GPS codes to measure the range between the orbiters and to determine the time offset between the two GPAs. The GPAs maintain internal time, not synchronized to external (spacecraft) time in order to expedite measurement of the precise time offset between the GPAs at the 100 nanosecond level. The accuracy and stability of this measurement is comparable to the stability of the GPS solution on GRACE. At power up each GPA starts a counter, and the units on each orbiter each compare their own LGRS time to that of the remote unit, and the unit with the more advanced time is taken to be the "master," while the unit with the lower time synchronizes its time to the master. This maintains continuous instrument time even after resets of either GPA.

3 Components of the LGRS

3.1 Ultra-Stable Oscillator (USO)

The USO provides three coherent RF outputs to other payload elements: a 38.656 MHz output as the sampling clock reference for the GPA; a 38.656 MHz output as the reference

Table 2 Key performance parameters of the USO. See Fig. 6 for Allan deviation

Parameter	Performance
Phase noise (dBc at 38.656 MHz)	-106@1 Hz
	-123@10 Hz
	-134@100 Hz
	-135@1 kHz
	-136@10 kHz
	-136@100 kHz
Temperature sensitivity (df/f per K)	5×10^{-13}
Aging rate (df/f per day)	$< 6 \times 10^{-11}$

Fig. 6 Pre-launch Allan deviations (ADEV) of each GRAIL flight USO measured against a hydrogen maser. ADEVs are indicative of each USO, except at and below 1 second where the maser may be the limiting factor

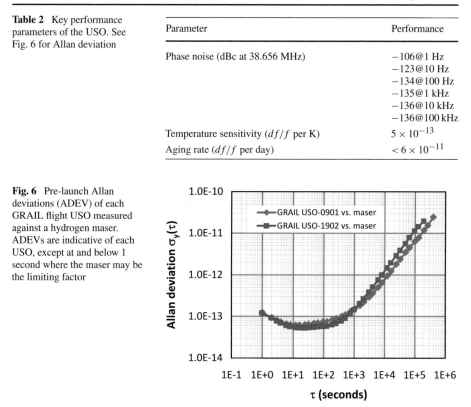

input to the RSB; and a 57.984 MHz reference input to the MWA. USOs on GRAIL-A differ in frequency from the USOs on GRAIL-B as shown in Table 1. Key performance parameters for the USO are shown in Table 2 and Fig. 6. The USO frequency must be calibrated from the ground periodically to prevent drifts in the USO frequency from corrupting the gravity measurements. This calibration occurs via tracking of the RSB signal by the DSN with a maximum interval between calibrations of 16 hours. Calibration data for both USOs is expected to be available when the DSN is in contact with either spacecraft, as the Radio Science Receiver (RSR) data will contain transmissions from both satellites simultaneously.

3.2 Gravity Recovery Processor Assembly (GPA)

The GPA is the central processor for measuring the baseband phase from the MWA and for generating and tracking the Time Transfer signal used to coordinate time between LGRS-A and LGRS-B. The timing information is required for the formation of the Dual One Way Range (DOWR) variable used to suppress the effect of USO noise on measurement of changes in the inter-satellite range. The Time Transfer Front End (TTFE) includes a diplexer to separate out the transmitted and received signals on the two orbiters and an amplifier for the received signal; the transmit and receive signals differ in frequency by 175 MHz. A helibowl antenna provides a minimum of 11 dBi gain. Critical performance parameters of the GPA and associated Time Transfer System (TTS) hardware are shown in Table 3.

The TTS provides the two-way time transfer link between the spacecraft by modulating GPS C/A ranging codes onto the LGRS S-band interspacecraft carrier signals. On GRACE

Table 3 GPA key performance metrics

Parameter	Value
Ka-band IF phase measurement error	$< 6 \times 10^{-6}$ cycles/Hz$^{1/2}$
Time offset measurement error	± 50 ns
Time offset measurement stability	< 632 ps/Hz$^{1/2}$
IF temperature sensitivity	< 0.002 radians/K
S-band transmit power	$+17$ dBm
S-band time transfer antenna gain	> 11 dBi within 5 degrees of boresight (3 dB half width ~ 20 degrees)
IF phase sensitivity to received amplitude	< 0.15 microns/dB
IF phase sensitivity to power supply voltage	< 0.13 microns/V

Table 4 Time Transfer System coding scheme

Item	Ebb	Flow
GPA Sample Rate (nom) MHz	19.328	19.328396
Divisor	20	19
Chips/cycle	1023	1023
Code length (s)	0.001058568	0.001005619
Code cycle ambiguity (km)	*317.3506669*	*301.4769568*
Cycles/databit	20	20
Databits/second	47.234	49.721
Message length (bits)	256	256
Prime factors	$2^6 \times 5^2 \times 151$	263×967
"Fortnight" (seconds)	1309440	1309440

GPS is used for this function. The coding scheme and timing are shown in Table 4. The code cycle ambiguities differ slightly between the two orbiters but are both approximately 1 millisecond, corresponding to approximately 300 km of range ambiguity. The full timing code sequence repeats every 1309400 seconds, slightly longer than a fortnight. Typical one-second voltage SNRs on the S-band measurements are approximately 2000 for ranges of 225 km. Performance of the S-band measurement easily supports the noise requirement as seen in Fig. 7.

The LGRS supports on-orbit software uploads, which were used in flight to enhance robustness and stability of the S-band tracking software, which was new to GRAIL.

3.3 Ka-band Ranging Assembly (KBR)

The KBR on each orbiter consists of a Ka horn assembly (KaA), a Microwave Assembly (MWA), waveguides connecting the KaA transmit and receive signals to the MWA, and the structure and thermal hardware supporting these elements.

The Microwave Assembly (MWA) converts the frequency reference signal produced by the USO to the Ka-band frequency which is transmitted to the other satellite and performs a direct, quadrature, down-conversion of the signal received from the other spacecraft using this transmit signal. Table 5 shows the MWA key performance parameters.

Fig. 7 The time transfer measurement stability easily supports its noise requirement

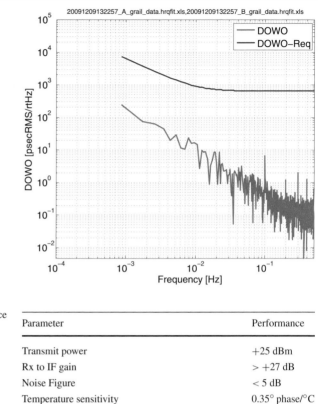

Table 5 MWA Key Performance Parameters

Parameter	Performance
Transmit power	+25 dBm
Rx to IF gain	> +27 dB
Noise Figure	< 5 dB
Temperature sensitivity	0.35° phase/°C

The Ka-band antenna provides a 26 dBi radiation pattern at 32 GHz designed to have high (50 dB) side-lobe suppression to mitigate multipath errors resulting from receipt of reflected rather than direct signals. It uses two orthogonal linear polarizations to isolate the transmit signal from the receive signal. The antenna is comprised of a horn and orthomode transducer. As on GRACE the polarization axes for transmit and receive are at 45 degrees to the spacecraft axis, so when the orbiters point at each other the polarizations align transmit from one orbiter into the receive of the other. A two-layer radome keeps light from the sun from entering the horn aperture to help maintain thermal stability. The use of two layers spaced by a quarter wavelength at Ka-band reduces reflections to help minimize sensitivity to thermal changes. Figure 8 shows the Ka horn antenna pattern.

3.4 Radioscience Beacon (RSB)

DSN tracking of the RSB allows calibration of the frequency of the USO, with a maximum time between calibrations of each USO of 16 hours under nominal tracking conditions. The RSB takes a ~ 38.656 MHz reference signal and synthesizes an X-band signal which is transmitted to the DSN by way of one of two identical antennas; exact frequencies are shown in Table 1. The antennas are on opposite faces of the spacecraft to provide full sky coverage, and during the mission are switched along with the spacecraft telecommunication antennas. The RSB on GRAIL-A and GRAIL-B operate at slightly different frequencies and can be recorded simultaneously by the Radio Science Receiver, which provides open-loop sampling

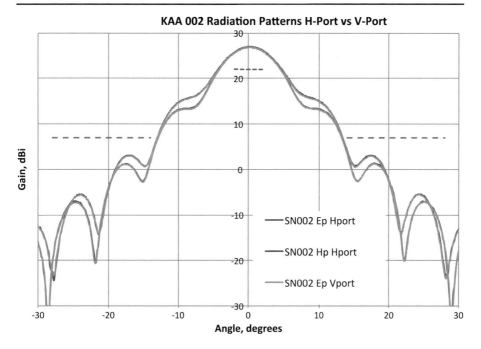

Fig. 8 The GRAIL KAA horn pattern achieves high gain and strong side lobe suppression to avoid noise from reflections off the lunar surface

of the received signal at a minimum rate of 100 kSamples/s for post-processing. One-way RSB tracking data will also be used in the science data analysis to provide enhanced Doppler data.

The primary requirements on the RSB are frequency stability relative to the USO input (easily met), and the Equivalent Isotropic Radiation Pattern, which is met by a combination of a +21 dBm transmit power and the gain pattern shown in Fig. 9.

4 Error Discussion

Two categories of instrument errors were considered: stochastic noise, which characterizes the sensitivity of the instrument; and deterministic errors, which were characterized as amplitudes at twice per orbit. These are discussed separately below.

4.1 Stochastic Noise

The LGRS measures a time series of separations, with a noise error budget naturally described as a Root Power Spectral Density (RPSD) of phase (range) fluctuations. Because the science analysis simulated the derived Range Rate data product, the requirements on the LGRS were translated from range to range rate $[S_{RR}(f) = (2\pi f)^2 S_R(f)]$, as shown in Fig. 10. For Fourier frequencies above 0.03 Hz the measurement noise is white phase (range) noise, visible in the slope of f in the "LGRS analysis" curve in Fig. 10. Key contributors to the noise include Ka-band link budget noise, GPA measurement noise, USO phase noise, and sampling time errors; these are shown as the four black and brown curves listed in Fig. 10. This analysis follows closely to the GRACE analysis (Thomas 1999).

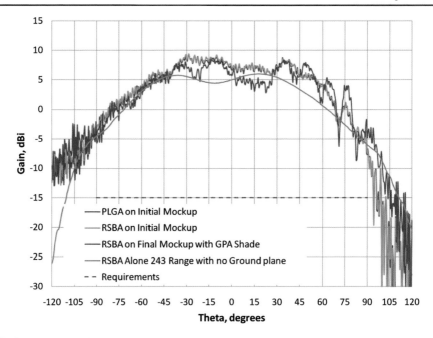

Fig. 9 Gain pattern of the RSB X-band patch antenna

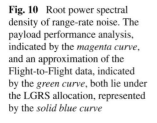

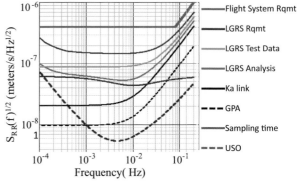

Fig. 10 Root power spectral density of range-rate noise. The payload performance analysis, indicated by the *magenta curve*, and an approximation of the Flight-to-Flight data, indicated by the *green curve*, both lie under the LGRS allocation, represented by the *solid blue curve*

Errors are dominated by the Ka-band link noise, which depends upon the spacecraft separation, the Ka horn gain, the MWA transmit power, system and processing losses, and the noise figure of the MWA receiver, all described above. These allow calculation of the phase readout noise as

$$\delta\mathrm{DOWR}_{1\text{-}way} = \lambda/(2\pi\,\mathrm{SNR}_V),$$

where SNR_V is the voltage signal-to-noise ratio for a given measurement bandwidth. Table 6 provides a sample of the parameters used in calculating SNR_V. This calculated sensitivity agrees reasonably with the observed two-way sensitivity of 0.5 microns/Hz$^{1/2}$ for a post-launch checkout described below when scaled for the 500 km separation during that test. Extensive testing was done pre-launch with signal levels comparable to separations between 50 km and 250 km, covering the expected mission parameters.

Table 6 Sample parameters used in the Ka-band link budget

Item	Value	Unit
Frequency	32.7	GHz
Transmitted power	25	dBm
Antenna gain	25	dBi
Separation	225	km
Noise temperature	550	K
Noise bandwidth	10	MHz
Noise power	−171	dBm/Hz
Received power	−96	dBm
Processing loss	3	dB
C/n_0	72	dB-Hz
SNR_V	5900	Peak sig V/(V_{RMS} Hz$^{-1/2}$)
One-way ranging noise	0.25	microns/Hz$^{1/2}$

Phase measurement by the GPA adds a small amount of noise dependent on the signal amplitude. Approximately 0.06 microns/Hz$^{1/2}$ was measured by testing the GPA with a synthesized 670 kHz IF signal representative of a 250 km separation. The synthesized IF presents a low-phase-noise signal to the GPA compared to the downconverted Ka-band signal, which contains both excess phase noise and a higher noise floor than the GPA itself.

USO noise enters several different ways. Fluctuations in the phase of the USOs look like fluctuations in the range, so the one-way data will have a power spectrum of range fluctuations, $S_{r12}(f)$,

$$S_{r12}(f) = S_\phi(f) \times \lambda^2/4\pi^2,$$

where $S_\phi(f)$ is the USO phase noise multiplied up to Ka-band, and $\lambda = 0.0092$ meters is the wavelength of the microwave signal. Since the MWA downconverts the incoming signal with its transmitted one, phase noise is correlated between the two orbiters. The DOWR is the combination of data from the two orbiters that removes this common mode phase noise in post processing. Following Thomas (1999), the DOWR combination suppresses noise with a gain factor, $G_{DOWR}(f)$ as

$$G_{DOWR}(f) = \frac{1}{4}\left[\left(\frac{\Delta f}{f_0}\right)^2 + \left(2\pi f \frac{r_{12}}{c}\right)^2\right],$$

where $\Delta f = 670$ kHz is the Ka IF frequency, $f_0 = 32.7$ GHz is the Ka-band frequency, r_{12} is the separation between the two orbiters, and c is the speed of light. The DOWR filter amounts to a high pass filter with the corner frequency at the light-travel time between the orbiters. Limits to this filtering come from the IF frequency, nominally 670 kHz for GRAIL.

The USOs drive the GPA samplers used to measure the Ka IF phase, and variations in the sampling time are indistinguishable from fluctuations in this phase. The ranging error spectrum from jitter can be written,

$$S_{jitter}^{\frac{1}{2}}(f) = S_{\delta t}^{\frac{1}{2}}(f) x \frac{670,000 \text{ cycles}}{s},$$

so a jitter in the sampling time of 632 picoseconds/Hz$^{1/2}$ would correspond to approximately 4 microns/Hz$^{1/2}$. This value was used as a performance requirement for the S-band measurement for times longer than 100 s, relaxed significantly compared to the TTS phase measurement noise floor of approximately 0.1 picoseconds. For a given amount of USO

Table 7 Temperature sensitivity at the unit level. Errors assumed worst case thermal variations. Items in bold are recommended for calibration using on-board temperature sensors

Unit	Unit performance	Verified by	ΔDOWR microns	$\delta\Delta$DOWR (microns)
USO	$< 4.5 \times 10^{-13}/°C$	Test	0.06 (@250 km)	0.06
GPA	< 0.002 rad/°C (GPA + TTFE)	Test	0 ± 1.6	1.6
MWA	0.35 ± 0.08 ° phase/°C	Test	0.7 ± 0.15	0.35
Ka horn	10.52 microns/°C	Test and Analysis	1.38	0.69
Ka waveguide	5.45 microns/°C	Test and Analysis	0.72	0.38
Ka Ret. Loss	<0.07 microns/°C	Analysis	0.09	0.09
TOTAL			**4.55**	**3.17**

phase noise these timing errors are much smaller than the corresponding noise in the multiplied Ka-band signal, but these do not cancel in forming the DOWR observable. At times shorter than a few hundred seconds the USO noise is sufficiently low not to be a contributor, and for longer times the S-band cross link measures the USO relative noise for correction in post-processing with errors well below the requirements level. The combination of TTS measurement noise and USO noise can be seen in Figs. 14 and 16.

The TTS link is insensitive to common mode changes to the USO frequencies, which amount to a scale change in the measurement,

$$\delta r = r_{12}\frac{\delta f}{f}.$$

Calibration of even one USO through tracking of the RSB signal then limits the growth of this error term, which dominates the instrument error budget at low Fourier frequencies.

4.2 Deterministic Errors

GRAIL used the Science Data System from GRACE to perform simulations linking instrument and mission parameters to science return (Asmar et al. 2013). These simulations found that orbit-correlated deterministic errors at twice per orbit had the largest impact on science performance, particularly detection of the lunar inner core. Thermal fluctuations dominate the deterministic error budget, with sensitivities and error contributions summarized in Table 7. Thermal fluctuations were at a minimum for $\beta > 70°$ as there were no spacecraft eclipses, so these data should represent the cleanest measurements (β is the angle between the sun-spacecraft vector and the orbital plane). The errors in Table 7 assume the worst case thermal fluctuations based on orbit simulations. Note that three of the items are deemed large enough to merit calibration using on-board temperature sensor data: the MWA, Ka horn and Ka waveguides. The spacecraft temperature sensors provided readout stability and resolution of 0.25 K over relevant timescales.

4.2.1 Non-thermal Deterministic Errors

In addition to the thermal items above, three non-thermal orbit-correlated errors were included in the error analysis: phase variations with signal amplitude, transmitted wavefront variations with spacecraft pointing, and time offset errors coupled to interspacecraft velocity.

Separation between the two orbiters varies throughout the orbit, resulting in variations in received signal power at Ka-band correlated with the orbit. The MWA showed 1.8 microns/dB sensitivity, and the GPA showed another 1.0 microns/dB. The GPA sensitivity depended on the received signal power, but these combined errors are modest when coupled to the approximately 1 dB received power changes.

As the pointing of the Ka horn relative to the distant orbiter varies, small variations in the electrical phase will be experienced independent of the geometric "phase center" effects. The worst case phase variation within $3°$ from boresight was measured to be $0.05°$. During the mission pointing from one orbiter to the other was required to be within $0.1°$, significantly suppressing this error down to an estimated 0.04 microns.

Timing between data samples taken on the two orbiters must be known to allow formation of the DOWR as well as to avoid coupling of interspacecraft velocity to phase; this latter effect sets the stricter requirement. This can be seen through consideration of the measured phase difference between the two sides:

$$\Delta\Phi = f_{IF} \times \Delta T,$$

where $f_{IF} = 670$ kHz is the Ka-band IF frequency, and ΔT is time offset between measurements. We are not sensitive to this phase offset, but we are sensitive to variations:

$$\delta\Delta\Phi = dF \times \Delta T + f_{IF} \times \Delta T.$$

The second term above reflects the sampling timing errors discussed above. Doppler shifts of up to 7 m/s would couple with 50 ns timing errors to add errors of approximately 0.35 microns.

Performance of the time offset measurement of the GPAs was tested by synthesizing a 670 kHz signal common to two GPAs, ramping the frequency of the IF over 10 kHz–100 kHz and observing the measured change in the DOWR and solving for

$$\Delta T = \frac{2}{\lambda} \frac{(\mathrm{DOWR}_{f2} - \mathrm{DOWR}_{f1})}{f_2 - f_1}.$$

This technique was routinely applied during ground software testing to ensure proper performance of the time-offset measurements and during thermal vacuum testing of the GPA. This test was not possible after instrument integration because the IF frequency is then fixed by the USO and MWA.

5 Instrument Performance and Testing

5.1 Pre-delivery Testing

LGRS development benefited from early development of a system-level radiated testbed housed in a 60-foot anechoic chamber, with 250 km free-space attenuation simulated by the free space loss augmented with a combination of freestanding microwave absorber and cabled attenuators. On one end a spacecraft mockup housed engineering model GPAs and microwave horns in conjunction with a GRACE spare MWA; low-noise synthesizers were used in place of USOs. The other end housed the microwave horn in a partial spacecraft mockup on a precision linear translation stage, allowing calibration and phasing checks of the signals. (See Fig. 11.)

The end-to-end performance in Fig. 12 shows the Ka-band noise meeting requirements for the designed 250 km signal level.

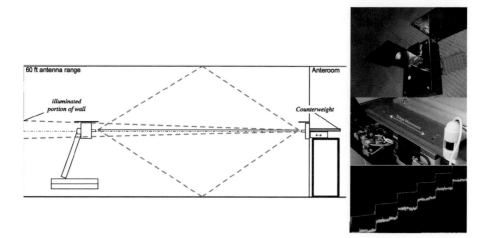

Fig. 11 LGRS engineering model hardware benefited from a radiated system-level testbed. One spacecraft mockup was on a precision translation stage for calibration and phasing tests. *Images at the right* show the movable mockup (*top*), the translation stage (*middle*) and commanded and measured response for 100 micron steps (*bottom*). *Dashed red lines* indicate the direct (inner) and multi-path (outer) RF signal paths

Fig. 12 Data from the radiated testbed demonstrates performance of the radiated link

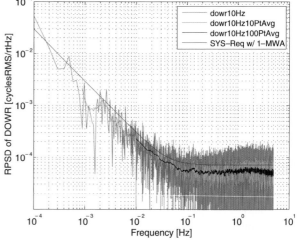

5.2 Flight-to-Flight Cabled Testing

Cabled electrical testing of the USO-MWA-GPA flight units destined for the two orbiters was used to verify radiometric performance prior to integration of the MWA into the KBR. This flight-to-flight testing augmented ongoing testing of each hardware complement against ground support equipment mimicking the second spacecraft.

The performance meets requirements for simulated separations between 50–250 km (see Fig. 13). In addition to the traditional thermal sensitivity of the cables, large variations in the DOWR were seen when the cables were mechanically perturbed. The spikes in the data causing the bump and spikes in the frequency spectrum around 10^{-3} Hz are strongly correlated with temperature measurements taken in the laboratory. These air conditioning cycles

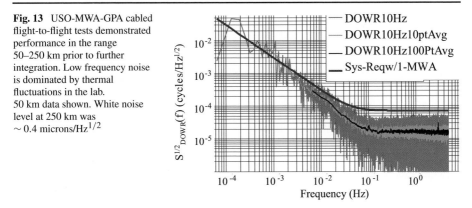

Fig. 13 USO-MWA-GPA cabled flight-to-flight tests demonstrated performance in the range 50–250 km prior to further integration. Low frequency noise is dominated by thermal fluctuations in the lab. 50 km data shown. White noise level at 250 km was ~0.4 microns/Hz$^{1/2}$

are clearly not representative of the flight environment and are considered challenges associated with the GSE. Since the deterministic errors, primarily thermally driven, are carried as a separate part of the error, these data runs are viewed as positive validation that the noise requirements are supported by these tests. The data easily supports the instrument performance requirements shown in Fig. 10.

Tests performed prior to spacecraft integration used hat couplers at S-band and Ka-band to allow cabling of the flight hardware to electronic ground support equipment (EGSE), which used engineering model and prototype hardware to mirror the complementary spacecraft. The white noise part of the spectrum could be observed, but radiated interference made performance verification difficult during this phase of integration and test.

5.3 Post-integration Testing

In addition to ongoing verification of the flight hardware against ground support equipment, a payload-to-payload demonstration was performed post integration to the spacecraft. During this test hat couplers on the S-band and Ka-band antennas were cabled together using coaxial cables. To avoid unintended radiated coupling between the antennas the test was performed with Flow in a shielded room providing electrical isolation. The purpose of this test was to demonstrate that the LGRS instruments on GRAIL-A (Ebb) and GRAIL-B (Flow) could track each other in their flight configuration on the orbiter. Specifically the GPAs on the orbiters were shown to acquire and track the Ka-band signals exchanged between the orbiters as the primary phase (biased range) measurement, and to acquire and track the S-band time transfer signal and synchronize to the "master."

While the test was intended only as a functional demonstration, the Ka-band data and S-band time offset measurements showed excellent performance as shown in Fig. 14. The root power spectral density (RPSD) of Ka-band phase fluctuations lies well below the requirement in the white noise section, which represents the radiometric performance. At low frequencies, variations in the cable electrical length rise above the requirement line as expected for long cables (this data corresponds to a fractional length stability of 4 parts per billion over 10 seconds).

The RPSD of the S-band time offsets, on the right half of Fig. 14, represents a measurement of the phase noise of the two flight USOs and matches the pre-integration performance. The USO noise is expected to cross the requirement line after approximately 300 seconds when the USO is fully warmed up (after approximately 2 days of continuous power). This data shows that the USO performs quite well after only a few hours of powered time.

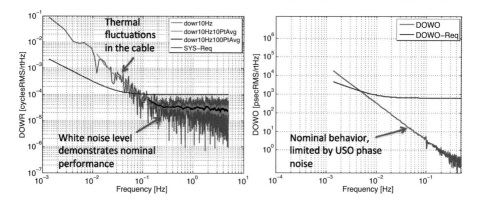

Fig. 14 Links were established successfully between the LGRS on GRA and GRB, demonstrating interoperability of the science instrument post integration to the spacecraft. The Ka-band tracking data (*left*) shows a white noise performance matching the expected instrument performance. Low frequency the noise crosses the requirement line as expected for tracking through long cables. Similarly the S-band data quality (*right*) matches expectations and easily supports the mission performance. Crossing of the requirement line at low frequencies is expected since the S-band measures real USO phase noise

An additional test was performed to test for interference of the S-band crosslink from the spacecraft telecommunication transmitter, also at S-band. This test was performed on Ebb, since the telecom transmitter frequency was relatively close to the LGRS S-band receive frequency. For this test Ebb was surrounded by a wall of portable microwave absorber and tested against ground support equipment. This test showed approximately 3 dB drop in the S-band SNR when the telecom transmitter was radiating, a level of interference easily supported by the instrument error budget. This degradation in S-band SNR on GRAIL-A (Ebb) has been observed in flight when the telecom system transmits through the antenna closer to the LGRS S-band antenna; the antennas are switched every two weeks to accommodate the changing geometry to Earth.

5.4 Post-launch Checkout

The LGRS met all of its performance requirements as verified prior to delivery. A post-launch checkout on 22 September 2011 allowed verification of in-flight performance when the orbiters were away from the strong "disturbances" in the science configuration at the Moon. The data shown in Fig. 15 demonstrate the sensitivity of the instrument even 1 million kilometers from Earth. Even at this distance, the gravitational influence of the earth results in a perceptible relative acceleration of the two spacecraft. The spectral content of this gravitational acceleration is compared to measurement on the left in Fig. 15. Filtering of this data to remove the "spectral leakage" from the Earth's acceleration allows the noise floor to be clearly seen meeting requirements (right side of Fig. 15). The demonstration was limited to approximately 20 minutes by thermal requirements of the spacecraft. Figure 16 also provides post-launch stability data of the USOs, among the best ever flown, and the performance of the S-band TTS.

5.5 Direct to Earth Time Correlation

The Science Data System must estimate the time offset between the LGRS data and the ground Doppler tracking used for orbit determination. Testing during spacecraft integration

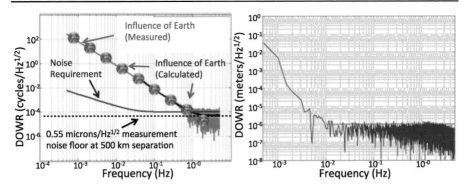

Fig. 15 Post-launch checkout shows excellent LGRS performance 1 million km from Earth. *Data on the right* has been filtered to demonstrate the instrument noise floor (see text)

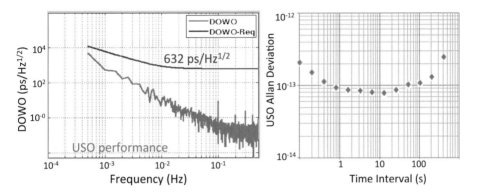

Fig. 16 Post-launch LGRS checkout demonstrates excellent time-transfer system measurement of the USO noise (*left*) and USO stability (*right*)

showed that the initial time offset knowledge requirement of 20 ms was easily met, but an additional in-flight test was used as an independent verification. In this Direct-to-Earth (DTE) test, the TTS signal was recorded with an RSR when the constellation was in a geometry that illuminated the Deep Space Network (DSN) Goldstone station, which occurred 3 times during the primary mission. Results of this test showed consistency of the spacecraft timing as described by Esterhuizen (2012).

5.6 In-flight USO Performance Verification

The instrument cross-links can be used to provide better measurement of USO performance in orbit than can be achieved from ground tracking. Distance fluctuations are measured as the sum of the Ka phases from the two spacecraft; taking the difference instead makes the link sensitive to clock noise and allows calculation of the Allan deviation (Allan 1966), a measurement of the stability between the two USOs. This Ka-band measurement has lower noise than the S-band link, which sets the synchronization of samples between the GPAs. As described in Enzer et al. (2012) evaluation of data post launch, en route to the Moon and at the start of science operations represent the first measurement of 10^{-13} level ADEVs from 1 to 100 seconds for USOs while in space.

6 Conclusion

The LGRS successfully adapted the measurement techniques used on GRACE to operation at the Moon on GRAIL. Performance of the LGRS has met its stringent measurement requirements, demonstrating sensitivity of 0.5 microns/Hz$^{1/2}$ at 500 kilometers separation, and down to approximately 0.2 microns/Hz$^{1/2}$ at 50 kilometers. It has been operationally robust, with only a few hundred seconds of lost data out of 90 days, for a total availability of 99.99 %. The resultant data quality coming from GRAIL lie in testament to strong performance and teaming among the payload, spacecraft, science analysis, and science teams.

Acknowledgements The GRAIL mission is supported by the NASA Discovery Program under contract to the Massachusetts Institute of Technology and the Jet Propulsion Laboratory, California Institute of Technology. The research described in this paper was carried out at JPL.

The authors have written this paper on behalf of the team of talented and hardworking engineers at JPL and at our contractor facilities. The success of the LGRS belongs to all of them.

References

D.W. Allan, Statistics of atomic frequency standards. Proc. IEEE **54**(2), 221–230 (1966)

S.W. Asmar et al., The scientific measurement system of the gravity recovery and interior laboratory (GRAIL) mission. Space Sci. Rev. (2013, this issue). doi:10.1007/s11214-013-9962-0

C. Dunn et al., The instrument on NASA's GRACE mission: augmentation of GPS to achieve unprecedented gravity field measurements, in *Proc. 15th Int. Tech. Meeting of Satellite Division of Institute of Navigation*, Portland, OR (2002), pp. 724–730

D.G. Enzer, R.T. Wang, K. Oudrhiri, W.M. Klipstein, In situ measurements of USO performance in space using the twin GRAIL spacecraft, in *Proceedings of the 2012 IEEE International Frequency Control Symposium* (2012), pp. 1–5

S. Esterhuizen, Moon-to-Earth: eavesdropping on the GRAIL inter-spacecraft time-transfer link using a large antenna and a software receiver, in *Proceedings of ION GNSS 2012*, Nashville, TN, USA, Sept 17–21, 2012

B.D. Tapley, S. Bettadpur, J.C. Ries, P.F. Thompson, M.M. Watkins, GRACE measurements of mass variability in the Earth system. Science **305**, 503–505 (2004). doi:10.1126/science.1099192

J.B. Thomas, An analysis of gravity-field estimation based on intersatellite dual-1-way biased ranging. JPL Report, Jet Propulsion Laboratory, Pasadena, CA (1999)

M.T. Zuber et al., Gravity recovery and interior laboratory (GRAIL): mapping the lunar interior from crust to core. Space Sci. Rev. (2013, this issue). doi:1007/s11214-012-9952-7